Capture and Relaxation in Self-Assembled Semiconductor Quantum Dots

The dot and its environment

Capture and Relaxation in Self-Assembled Semiconductor Quantum Dots

The dot and its environment

Robson Ferreira and Gérald Bastard

Laboratoire Pierre Aigrain Ecole Normale Supérieure,
24 rue Lhomond – 75005 – Paris

Morgan & Claypool Publishers

ISBN   978-1-6817-4089-8 (ebook)
ISBN   978-1-6817-4025-6 (print)
ISBN   978-1-6817-4217-5 (mobi)

DOI   10.1088/978-1-6817-4089-8

Version: 20151201

IOP Concise Physics
ISSN 2053-2571 (online)
ISSN 2054-7307 (print)

A Morgan & Claypool publication as part of IOP Concise Physics
Published by Morgan & Claypool Publishers, 40 Oak Drive, San Rafael, CA, 94903, USA

IOP Publishing, Temple Circus, Temple Way, Bristol BS1 6HG, UK

Contents

Preface   viii

Acknowledgements   ix

Author biography   x

1   Electronic states in self-assembled semiconductor quantum dots   1-1

1.1   Simple image of a self-assembled semiconductor quantum dot   1-1

    1.1.1 Issues in the modelling of quantum dots   1-1

1.2   Quantum dot bound states   1-4

    1.2.1 The effective Hamiltonian and its symmetry   1-6

    1.2.2 The simple variational estimate   1-7

    1.2.3 A quasi-separable approach   1-8

    1.2.4 Exact numerical calculations   1-9

    1.2.5 Examples of variational energies and wavefunctions   1-11

1.3   Continuum states   1-14

1.4   Coulombic interactions   1-15

    1.4.1 Two-electron states   1-15

    1.4.2 Electron–hole pairs   1-19

    1.4.3 Multi-charged dots   1-20

1.5   Phonons and polarons in InAs quantum dots   1-20

    1.5.1 Modelling the far-infrared absorption: the 'artificial atom'   1-22
        picture

    1.5.2 Far-infrared absorption at zero field and anisotropic   1-23
        macro-atom model

    1.5.3 Far-infrared magneto-absorption and breakdown   1-25
        of the macro-atom model

    1.5.4 Qualitative evidence for longitudinal optical phonons effects 1-26

    1.5.5 Phonons in bulk III–V materials: two essential features   1-26

    1.5.6 Fröhlich coupling between electrons and longitudinal   1-28
        optical phonons in bulk

    1.5.7 Phonons and Fröhlich interaction in a dot: a few assumptions   1-30

    1.5.8 Non-perturbative Fröhlich coupling in quantum dots   1-31

    1.5.9 Optical absorption by quantum dot polarons   1-34

    1.5.10 Polaron features in a quantum dot ensemble   1-36

    1.5.11 Analytical model for the quantum dot polarons   1-37

1.5.12 Localized phonon mode   1-43

1.5.13 Dynamical aspects of the polaron states   1-44

References   1-45

2   Capture of carriers by the quantum dots   2-1

2.1   Phonon-assisted capture by one empty dot   2-2

      2.1.1 Energy and size selectivity   2-2

      2.1.2 Effects of temperature and gas density   2-4

      2.1.3 Simple analytical model for the capture   2-7

2.2   Phonon-assisted capture by a charged dot   2-11

      2.2.1 Two opposite charges   2-12

      2.2.2 Two identical charges   2-14

2.3   Coulombic-assisted capture by one empty dot   2-16

      2.3.1 Scattering rate and indistinguishability effects   2-17

      2.3.2 Gas density dependence of the Auger capture   2-18

      2.3.3 A simple model for the Auger capture coefficients   2-20

      2.3.4 Auger capture of a second carrier   2-22

2.4   Conclusion   2-22

      References   2-23

3   Energy relaxation of confined carriers in self-assembled   3-1
      quantum dots

3.1   The phonon bottleneck in quantum dots   3-2

3.2   Polaron coupling versus phonon anharmonicity   3-7

3.3   The existence of energy windows associated with anharmonic decay 3-8

3.4   Modelling of unstable polarons: main issues   3-10

3.5   General approach to the unstable polaron states   3-11

      3.5.1 Stationary polarons   3-13

      3.5.2 Semiclassical relaxation model   3-13

      3.5.3 Polaron relaxation model   3-17

      3.5.4 Comparison of polaron and semiclassical predictions   3-18

      3.5.5 Experimental evidence for many-channel relaxation   3-19

3.6   Many-channel relaxation of quantum dot polarons   3-20

      3.6.1 Relaxation of high-energy polarons   3-21

      3.6.2 Relaxation of low-energy polarons   3-24

      3.6.3 Conclusion   3-25

3.7   Intra-dot Auger relaxation   3-26
    3.7.1 Two electrons   3-28
    3.7.2 One electron–hole pair   3-29
    3.7.3 Other (temperature-dependent) intra-dot relaxation processes   3-29
    References   3-34

Preface

Quantum dots are among the most studied nano-objects in semiconductor physics. They are characterized by a small number of strongly bound states, separated from a continuum of high-energy states with extended wavefunctions, recalling thereby an atomic series of levels. However, this 'macro-atom' picture, widely used in the literature, presents severe limitations. In particular, the idea that there exists a sharp separation between the quantum dot and its environment is very tempting but it has been shown to be highly misleading. Indeed, in many situations of interest, a population and/or energy exchange between the dot and its environment is a crucial issue. Many interesting applications of dots are based on these exchange processes. An archetypal example of coupling between the dot and its environment is the quantum dot laser, where electrons and holes injected in the system (either optically or electrically) should be captured by the dot and relax down to their ground states before participating in the lasing process. Dot-based photodetectors also require an efficient recapture process, to re-initialize the dot after its ionization by a photo-absorption process. As a final example, let us mention the single-photon emitters where an intense optical pulse is used to generate unbound electron–hole pairs: the carriers capture by the dot and energy relaxations are subsequent intermediate steps before the desired last photon emission occurs.

The present book results from a series of lectures given by the authors at the Master level. It presents an overview of different models and mechanisms developed to describe capture and relaxation of carriers in quantum dot systems. Despite their undisputed importance, the mechanisms leading to population and energy exchanges between a quantum dot and its environment are not yet fully understood. We develop here a first-order approach to understanding such effects. The methods employed are mainly those of elementary quantum mechanics. An introduction to the physics of semiconductors (notions of energy bands, effective mass and crystal lattice vibrations: phonons) is desirable.

The origin of the book also explains its unusual form. The lectures were delivered in the form of slides, often distributed to the students accompanied by short texts recalling the main points of the panel contents. Here, the discussion has been considerably enlarged and enriched with additional material. It is hoped the reader will find the format of the book convenient and helpful.

It is finally worth stressing the impressive amount of experimental work that exists on the physics and applications of self-assembled semiconductor quantum dots in general, and about the topics of capture and energy relaxation in particular. In this book, we chose to emphasize modelling issues and present only a very few experimental results, even though the remarkable advances of recent decades result from the interplay of both such aspects.

Paris, October 2015

Acknowledgements

An important part of the material presented in this book results from works of PhD students or joint research projects, and we have benefited from fruitful discussions with many colleagues. The authors acknowledge Drs Y Arakawa, U Bockelmann, G Cassabois, C Delalande, L-A De Vaulchier, C Diederichs, J-M Gérard, Th Grange, Y Guldner, S Hameau, K Hirakawa, T Inoshita, J-N Isaia, C Kammerer, Ph Lelong, I Magnusdottir, J-Y Marzin, E Molinari, N D Phuong, V Preisler, N Regnault, F Rossi, Ph Roussignol, H Sakaki, M Skolnick, G Strasser, K Unterrainer, A Vasanelli, O Verzelen, J Wang and L Wilson.

Besides the individuals listed below with whom we had pleasure in interacting, one of us (GB) would like to acknowledge the Technical University Wien, the Institute of Industrial Science, Tokyo University, and the Physics Department of the Hong Kong University of Science and Technology for their long term support.

Author biography

Gérald Bastard

Gérald Bastard received a PhD thesis from Paris 7 University on 'Magneto-optical investigations of $Hg_{1-x}Mn_xTe$ alloys' in 1979. He was post-doctoral fellow at IBM between 1981 and 1982 in Leo Esaki's group and regularly visits the Technical University Vienna, Institute of Industrial Science Tokyo, Hong Kong University of Science and Technology and Lund University. Dr Bastard is currently Directeur de Recherche CNRS (emeritus). He has written a textbook on 'Wave mechanics applied to semiconductor heterostructures'. Gérald Bastard is a fellow of the American Physical Society (1993) and was awarded several prizes such as the Fujitsu Quantum Device Award (2000) and the Heinrich Welker prize for compound semiconductors (2014).

Robson Ferreira

Robson Ferreira received a PhD from the Pierre et Marie Curie (Paris 6) University in 1992. He is presently Directeur de Recherche at the French Council for Research (CNRS), working on the electronic states and optical properties of low dimensional materials and heterostructures. He has additionally been involved with teaching, lecturing for many years on semiconductor materials and their nanostructures (Master programs at Paris 6 University and École Normale Supérieure-Paris).

Capture and Relaxation in Self-Assembled
Semiconductor Quantum Dots
The dot and its environment
Robson Ferreira and Gérald Bastard

Chapter 1

Electronic states in self-assembled semiconductor quantum dots

1.1 Simple image of a self-assembled semiconductor quantum dot

The quantum dots (QDs) of our interest are those obtained by the self-organized growth technique (see figure 1.1). In this technique, one starts by sending atomic fluxes of In and As on a thick GaAs substrate. Despite the huge lattice–period mismatch between the two compounds ($\Delta a/a_{GaAs} = 7\%$, where $a_{GaAs} = 5.65$ Å and $a_{InAs} = 6.06$ Å), the growth is initially bi-dimensional (2D); that is to say, an In plane grows, followed by an As plane, leading to a very thin layer of InAs with GaAs lattice parameter in the layer plane: its formation introduces a distortion of the atomic orbitals and thus excess energy in the chemical bonds. However, beyond a critical InAs thickness ($d \approx 0.7$ monolayer, i.e. before the complete growth of one monolayer) the elastic energy accumulated by the distorted orbital bonds becomes too large. The nature of the growth changes spontaneously (thus the term 'self-organized') to becoming three-dimensional (3D), which leads to the formation of unstrained (or less strained) InAs islands floating on a strained InAs 'wetting' layer (WL). Figure 1.1 shows an atomic force microscope image obtained *in situ* (i.e. inside the growth chamber) after the InAs islands have been formed and before they are buried by subsequent deposition of GaAs. This figure reveals that the self-organized growth results in an inhomogeneous distribution of islands with an average areal density of 10^{10}–10^{11} cm^{-2}. Thus, islands are fairly separated from each other and in the following we shall consider them as isolated QDs.

1.1.1 Issues in the modelling of quantum dots

Figure 1.1 does not allow precise information on the island shape. However, many studies (by imaging techniques, as in figure 1.1) or via indirect evaluations) have pointed out that these objects are very flat (height $\ll$ basis dimensions: for the sake

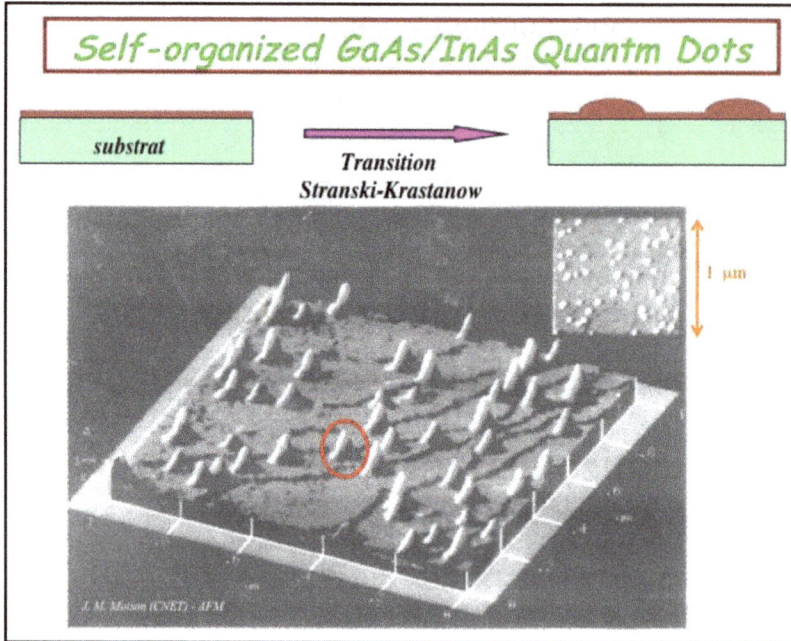

Figure 1.1. Atomic force microscope picture of InAs/GaAs self-organized quantum dots. Top: a schematic of the Stranski–Krastanov growth mode. Below: a top view image. (Courtesy of J M Moison.)

of visualisation of the dots, the vertical scale was dilated in the figure). In the following, we shall usually model them as cones (truncated or not) with height h and basis radius R, floating on a thin InAs WL ($d \ll h < R$ in figure 1.2). However, some of the results of calculations to be presented were obtained using different (lens-like, pyramidal) shapes. Actually, the modelling of quantum dots is quite complex for several reasons. Firstly there is *de facto* a lack of precise sizes and shapes. This arises from the fact that characterization techniques are often destructive or are performed on samples that are different from those that have been used in, say, optical measurements. It is well established that the dots' parameters crucially depend on the growth conditions. For instance, the (Ga,In)As dots are bigger than the InAs ones. Moreover, one faces severe diffusion problems (e.g. In has a tendency to move towards the surface during growth) that generate non-trivial composition profiles inside the islands. Other parameters, such as the substrate temperature, also play an important part. To summarize, a sample like that in figure 1.1 displays an inhomogeneous ensemble of dots whose average characteristics are functions of the growth parameters. Recent studies, in particular those undertaken on single QDs, show that precise knowledge of these growth parameters (composition gradient, geometrical shape, strain state of the QD vicinity, etc) may prove mandatory to explain the fine structure of the electronic states in semiconductor QDs.

Let us delve into the latter issues in more detail. Indeed, it is important to realize that the reliability of energy level estimates in dots is more dependent on our

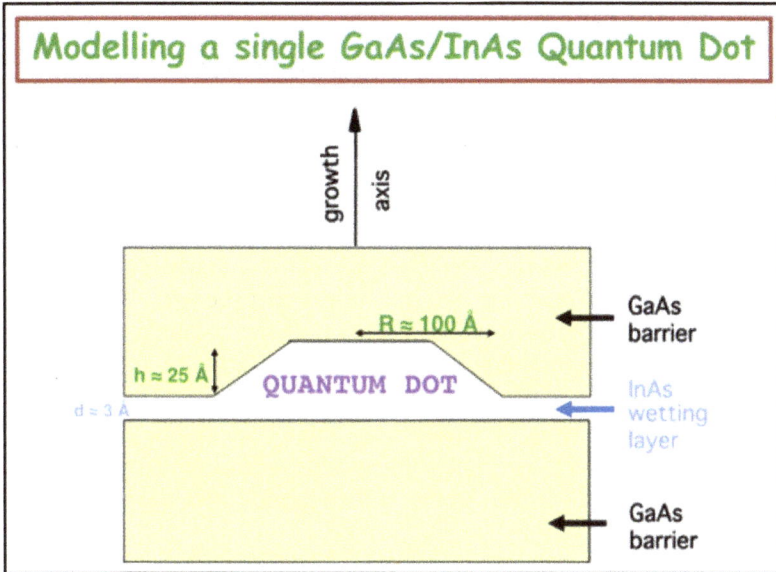

Figure 1.2. Sketch of a truncated cone InAs/GaAs QD floating on a thin InAs 'wetting' layer and surrounded by thick GaAs barriers.

knowledge of what the dot is in reality than a matter of technical difficulty of numerical analysis. In fact, a dot is a nanometric object: for instance, a (Ga,In)As dot self-organized in a GaAs matrix is a flat object with a height of a few nanometres and a radius typically around or bigger than 10 nm. Hence, electrons are in a marked quantum regime where the confinement energies sensitively depend on actual sizes. But how much do we know about those sizes for a particular dot? Very little indeed: in experiments that involve a single dot, the dot under investigation has most likely been selected at random and since self-organized dots have a distribution of sizes, the size of the particular dot under consideration is unclear. Also, the actual dot profile is not known: the perfect truncated cone in figure 1.2, used in various model calculations, is unlikely to be realized in actual samples. Other factors that remain unclear include the Ga–In interdiffusion and strain potentials in the dot, and the electrostatic potential fluctuation in the immediate vicinity of that particular dot (whether there is a trap nearby that randomly loads and discharges electrons/holes and leads to fluctuating Stark shifts of the dot energy levels). In experiments that involve ensembles of dots, different sizes and profiles show up as well as different sorts of dot environments. Hence, only an 'averaged property' can be extracted from experiments in a dot ensemble. In the specific example of optical experiments, the emission and absorption spectra display important linewidths (typically some tens of milli-electron volts), reflecting the inhomogeneous distribution (Marzin *et al* (1994)), whereas their peaks give a rough estimate of the characteristics (in particular the size) of the most numerous dots. (For a discussion of the fundamentals of QDs, see the contribution of Bastard and Molinari in Rossi and Zanardi (1995) and references cited therein.)

1.2 Quantum dot bound states

Suppose now that we know/assume the shape/environment of the dot (such as the one in figure 1.2, for example). There are two classes of energy level calculations. The 'atomistic' calculations can be termed empirical *ab initio*. One takes the pseudopotentials of the different atoms (that one knows from suitable adjustments from bulk semiconductors) and then one solves a huge matrix that has at least the dimension of the number of atoms in the dot (plus those in the vicinity, since the dot is inserted in a solid-state matrix). Another 'atomistic' calculation takes an sp^{3}s* (or sp^{3}s*d) atomic parameterization of the dot atoms and the surrounding atoms, and solves the tight binding problem assuming (or not) nearest neighbour coupling (in other words, each As is surrounded by four Ga atoms in GaAs and four In atoms in InAs). Again, one ends up with very large matrix diagonalization. These 'atomistic' methods are more ideally suited to handle very small dots such as the tiny CdSe, CdS colloidal dots (obtained by a very different growth technique, and of much smaller sizes; typically 1–2 nm in all directions), which means that they look more like a molecule than a solid.

Obviously, 'atomistic' methods face numerical difficulties when the dot dimensions increase. Then, a continuous description of the electronic states in the dot may be better suited. Such a description is provided by the envelope function approximation, where an effective Hamiltonian that acts on slowly varying envelopes takes the effect of the potential energies inside and outside the dot into account. The effective Hamiltonian can be a scalar (single band) or a matrix whose dimension is the number of bands that one wants to retain. Valence band (holes) states usually require a 4×4 effective Hamiltonian. The effective Hamiltonian also takes into account stress and strain effects. We shall not attempt to rigorously justify its pertinence for modelling the QDs, but only mention that it has been successfully utilized to study many of the dot properties we shall consider in this book. For a more detailed discussion of the envelope function method applied to different semiconductor heterostructures see, for example, Bastard (1988), Bastard *et al* (1991), Bimberg *et al* (1999) and Stier (2001).

In the following, we restrict ourselves to the simplest approach, namely a single band description. In this framework, all we need to know is the electron (or hole) effective masses inside and outside the dot (let us call them m_{dot}, $m_{barrier}$) and the confining potential. The Hamiltonian we study is therefore

$$H_{QD} = \vec{p}\left(\frac{1}{2m(\vec{r})}\right)\vec{p} + V_{QD}(\vec{r})$$

$$m(\vec{r}) = \begin{cases} m_{dot} & \text{if } \vec{r} \in \text{dot} \\ m_{barrier} & \text{if } \vec{r} \in \text{barrier,} \end{cases} \tag{1.1a}$$

where $V_{QD}(\vec{r})$ is the confinement potential, which may still depend on fluctuations in Ga and In composition and on strain profile. In the following, we will very often assume a position-independent effective mass m^* (i.e. the same mass in the dot, WL and barrier regions). Thus,

$$H_{QD} = \frac{\vec{p}^2}{2m^*} + V_{QD}(\vec{r}). \tag{1.1b}$$

Finally, in the crudest approximation one neglects any composition fluctuations and strain effects, so that V_{QD} is piecewise constant, i.e. it has a fixed value in the region occupied by the QD and in the WL, and another fixed value in the surrounding GaAs. A sketch of a truncated cone QD is shown in figure 1.2. Therefore, the GaAs material acts as a potential barrier for electrons in both the conduction and valence bands.

Under such assumptions the QD problem to be solved simplifies into one we could find in standard quantum mechanics textbooks (see, for example, Schiff (1968)). Also, as discussed in many of them, the dominant trend in the presence of an attractive potential is the existence of localized quantum states: in quantum dots, bound states follow from the fact that the InAs region has a much larger attractivity for electrons than the outside GaAs ones, and that the typical sizes of the QDs are at the nanometre scale.

It is however important to stress that even for such a simplified model for the confining potential, the calculations of the energy levels and their associated wavefunctions are far from trivial. In fact, the potential energy is 3D, deep and has short range. Also, as discussed in many textbooks, there are several methods that can be employed to find the eigenstates of H_{QD}. Let us quote two of them.

- Quick variational ansätze provide an idea of the confinement energies at small numerical expense, in general for the low-lying highly localized states. They are approximate and their accuracy depends on the choice of the variational wavefunctions. They may run into trouble when the confinement is marginal and they become awkward when tackling the high-energy continuum states.
- Exact numerical solutions are obtained by projecting H_{QD} on a suitable basis. This basis is usually the solution of the free particle Hamiltonian inside a prescribed box. The latter should have dimensions that are large compared to those of the dot (to prevent edge effects). Its shape is *a priori* arbitrary, but using a shape that complies with the symmetry displayed by H saves a lot of numerical computation time. For instance, in the case of (Ga,In)As dots, the shape suggests that V_{QD} is approximately invariant under rotation around the z-axis. Thus, a convenient large box will be a circular cylinder with radius $R_{box} \gg R$ (typically $R_{box}/R \approx 10$) and a height $H \gg h$ (see figure 1.3). In the case of a spherical dot, the confining box would be a sphere with $R_{box} \gg R$. In the case of dots with an awkward geometry, a cubic box appears reasonable (for example, for a CdSe tetrapod studied in Pang *et al* (2005), not discussed in this book, a cube edge parallel to one of the four legs of the tetrapod appears suitable).

In the following we apply both of the two mentioned methods and also present a third 'mixed' one to tackle the energy eigenstates of our QDs.

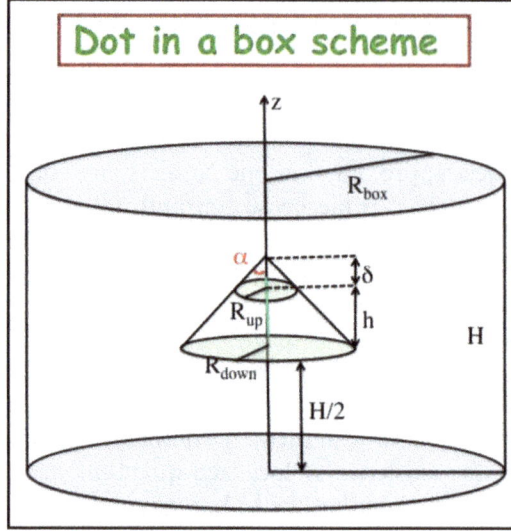

Figure 1.3. Parameters for a truncated cone embedded in a large cylindrical box. The WL (not drawn for the sake of clarity) lies underneath the dot: $-d \leqslant z \leqslant 0$ when $z = 0$ is taken at the basal plane of the cone. We often use the abbreviation $R_{\text{down}} = R$ in the following.

1.2.1 The effective Hamiltonian and its symmetry

For any numerical method, one gains in exploiting the dot shape symmetries. For a dot that displays cylindrical symmetry along the Oz-axis (the growth axis in figures 1.2 and 1.3), it is easy to show that $[H_{\text{QD}}, L_z] = 0$, where L_z is the projection of the angular momentum operator along Oz. Hence, we know that we can search the eigenvalues of H among the L_z ones. The latter are easy to find since in cylindrical coordinates

$$x = \rho \cos \varphi, \quad y = \rho \sin \varphi$$

$$\frac{\partial^2 \psi}{\partial x^2} + \frac{\partial^2 \psi}{\partial y^2} = \frac{1}{\rho} \frac{\partial}{\partial \rho} \left(\rho \frac{\partial \psi}{\partial \rho} \right) + \frac{1}{\rho^2} \frac{\partial^2 \psi}{\partial \varphi^2} \tag{1.2}$$

$$L_z = x p_y - y p_x = -i\hbar \frac{\partial}{\partial \varphi}.$$

Hence, the eigenfunctions of L_z are $\exp\{im\varphi\}/(2\pi)^{1/2}$ with the eigenvalues $m\hbar$ and $m = 0, \pm 1, \pm 2, \ldots$ The fact that m is a relative integer comes from the fact that if we change φ by 2π the wavefunction should be unchanged. $m = 0$ states are called S states, $m = \pm 1$ states are called P states, $m = \pm 2$ states are called D states, etc. Note that this same notation applies to a flat (2D) hydrogen atom and differs from the one for a regular 3D atom. Both atoms possess a single fundamental orbital ground state, named 1S in either case, but their spectra differ for the excited shells. In particular, there are only two 1P ($m = \pm 1$) and two 1D ($m = \pm 2$) states in 2D, while 3D spherically symmetric atoms have three 1P states and five 1D states. It is also worth pointing out two important specificities of the dots as compared to the 2D and 3D H atoms:

- The intermediate character of dots, in between 2D and 3D confining systems. Indeed, the vertical motion is not completely suppressed, as it is in 2D hydrogen, even though the bound dot orbits are much more confined along Oz than in the plane. This has important consequences in big dots, where for a given radial symmetry one can have more than one bound orbit (for instance, two 1S states) that differ in regard to their vertical motions. However, for most of the dots we shall consider in this work, the dot heights are quite small (1–2 nm), and excitations along the Oz direction are usually not present, leading to a 2D H atom-like labelling of the bound levels. Note also that radial excitations for a fixed m are also possible in a QD; as they are for 2S excited states, for instance. Of course, for a dot with axial symmetry, the $|N, \pm|m|\rangle$ states are degenerate for a given N ($N = 1, 2,...$), like the 2P$_+$ ($N = 2$, $m = +2$) and 2P$_-$ ($N = 2$, $m = -2$) states, for example.
- Anisotropic dots. As mentioned above in relation to figure 1.1, the actual shape of dots is not known. The assumption of axial symmetry is only a first approximation. Even if it allows many useful aspects of the dots' energetics to be explained (such as the important confinement energies we shall discuss in the next paragraphs), many experiments (some of them considered in the next chapter) clearly show the existence of some in-plane anisotropy. To the lowest order, a weak deviation from rotational invariance may lift the twofold degeneracy of the cylindrical excited states. As we shall see in chapter 2, optical experiments suggest the existence of a D-like radial asymmetry; namely, that two orthogonal in-plane axes are slightly unequal. To be explicit, this is the case when the dot shape is slightly elongated (contracted) along, say, the Ox (the Oy) direction: its basal projection resembles that of a rugby ball and no longer of a soccer ball. In this case, the ground $m = 0$ state is not affected (to the lowest order), while the $m = \pm 1$ states mix and lead to the formation of two new states (we shall consider these effects in more detail in the next chapter).

In the rest of this chapter we focus on axially symmetric dots.

1.2.2 The simple variational estimate

For the ground state, we expect a nodeless wavefunction. Hence, we can use the trial normalized wavefunction

$$\psi(\rho, z) = \frac{1}{\beta\sqrt{\pi\xi\sqrt{\pi}}} \exp\left\{-\frac{\rho^2}{2\beta^2} - \frac{(z - z_0)^2}{2\xi^2}\right\}. \tag{1.3}$$

Its physical meaning is simple: the electron is localized radially as well as along the Oz-axis. The confinement extensions β and ξ and the position z_0 are parameters that we can attempt to optimize. We expect that $z_0 > 0$ as long as the WL thickness is significantly smaller than the dot height: $d \ll h$. One finds the expectation value of the kinetic energy operator $T = \vec{p}^2/(2m^*)$ to be

$$\langle T \rangle = \langle T_r \rangle + \langle T_z \rangle = \frac{\hbar^2}{2m^*}\left(\frac{1}{\beta^2} + \frac{1}{2\xi^2}\right). \tag{1.4}$$

For the evaluation of the expectation value of the potential energy one takes $V_{QD}(\vec{r}) = 0$ in the GaAs barriers and $V_{QD}(\vec{r}) = -V_b$ in the InAs regions (dot and WL) to give

$$\langle V_{QD} \rangle = \frac{-V_b}{\xi\sqrt{\pi}}\int_{-d}^{h} dz \; e^{-\frac{(z-z_0)^2}{\xi^2}}\left\{1 - e^{-\frac{R(z)^2}{\beta^2}}\right\}$$

$$\frac{R(z)}{R_{down}} = \begin{cases} \dfrac{h - z + \delta}{h + \delta} & \text{if } z \in \text{dot} \quad (0 < z < h) \\ \infty & \text{if } z \in \text{WL} \quad (-d < z < 0) \end{cases}$$

$$\delta = h\frac{R_{up}}{R_{down} - R_{up}}. \tag{1.5}$$

Note that the integrals in the last equation are expressible in terms of the error function, which is well behaved and tabulated. Then, the total energy is minimized with respect to the three variational parameters ξ, β and z_0.

1.2.3 A quasi-separable approach

A hybrid method between the simple variational ansatz and all-numerical computations is the separable approach. It is well suited for dots that display a pronounced aspect ratio: very long needles ($h \gg R$, not discussed in this book) or disc-like (Ga,In)As dots ($h \ll R$). The separable ansatz method consists of looking for approximate solutions that can be written as

$$\psi(\rho, \varphi, z) = e^{im\varphi} f(\rho)g(z), \qquad L_z|\psi\rangle = m\hbar|\psi\rangle, \tag{1.6}$$

where the rotational invariance has been taken into account. The factorization $f(\rho)g(z)$ is the separability ansatz. In addition, one can restrict the choice of $f(\rho)$ by imposing a prescribed shape, for example

$$f(\rho) = \frac{1}{\beta\sqrt{\pi}}\exp\left(-\frac{\rho^2}{2\beta^2}\right), \tag{1.7}$$

where β is a variational parameter. Then, one multiplies $H_{QD}|\psi\rangle = E|\psi\rangle$ on the left by $e^{im\varphi}f(\rho)$ and integrate over the in-plane variables, to end up with a 1D

Schrödinger equation, explicitly dependent upon β for the unknown function $g(z)$ in the presence of an effective (averaged) potential,

$$[T_z + \langle T_r \rangle + V_{\text{eff}}(z)]g(z) = E \, g(z)$$

$$V_{\text{eff}}(z) = \left\langle e^{im\varphi} f(\rho) \middle| V_{\text{QD}}(\vec{r}) \middle| e^{im\varphi} f(\rho) \right\rangle$$

$$\langle T_r \rangle = \langle e^{im\varphi} f(\rho) | T_r | e^{im\varphi} f(\rho) \rangle. \tag{1.8}$$

The latter is then solved numerically. Call ε_m the eigenvalues for the z motion. They are parametrized by β and according to the variational theorem one should look for the lowest eigenvalue, i.e. minimize $\varepsilon_m(\beta)$ with respect to β.

Let us apply this procedure to S states ($m = 0$). The in-plane kinetic energy is obtained readily: $\langle T_r \rangle = \hbar^2/(2m^*\beta^2)$. To find the β-dependent effective potential energy $V_{\text{eff}}(z)$ for the z motion, we average the actual dot potential energy $V_{\text{QD}}(\rho, z)$ over the radial motion to give

$$V_{\text{eff}}(z) = -V_b \, Y(z+d) \, Y(h-z) \left\{ 1 - \exp\left[-\frac{R(z)^2}{\beta^2} \right] \right\}, \tag{1.9}$$

where the z origin is at the bottom of the truncated cone, the energy zero at the conduction band edge of the barrier-acting material, and $R(z)$ is defined in equation (1.5). Then, $V_{\text{eff}}(z) = 0$ both underneath the WL ($z < -d$) and above the top of the dot ($z > h$), and $V_{\text{eff}}(z) = -V_b$ in the WL region ($-d \leqslant z \leqslant 0$). Finally, $V_{\text{eff}}(z)$ differs from a piecewise constant function only inside the dot region ($0 \leqslant z \leqslant h$). Next, we solve equation (1.8) numerically: for each value of β, we compute the eigenenergies $\varepsilon_m(\beta)$ and then minimize the ground solution with respect to β. A similar procedure applies for the $m \neq 0$ states (see section 1.2.5 below).

1.2.4 Exact numerical calculations

In the numerical approach the quantum dot is placed at the centre of a large cylindrical box (see figure 1.3). The total Hamiltonian now is $H_{\text{QD+box}} = H_{\text{QD}} + V_{\text{box}}(\vec{r})$. We seek first the eigenstates of a free particle in the large box without a dot, i.e. we have to find the solutions of $H_{\text{QD+box}} - V_{\text{QD}}(\vec{r})$, namely

$$-\frac{\hbar^2}{2m^*}\left[\frac{1}{\rho}\frac{\partial}{\partial \rho}\left(\rho \frac{\partial \psi}{\partial \rho} \right) + \frac{1}{\rho^2}\frac{\partial^2 \psi}{\partial \varphi^2} + \frac{\partial^2 \psi}{\partial z^2} \right] = \varepsilon \psi, \tag{1.10}$$

with ψ vanishing at the box boundaries $z = \pm H/2$, and $\rho = R_{\text{box}}$, since $V_{\text{box}}(\vec{r})$ describes an impenetrable box ($\vec{r} = 0$ is the common centre of the basal dot surface and of the big box). The functions ψ factorize to yield

$$\psi(\rho, \varphi, z) = N \, e^{im\varphi} f_m(\rho) \sin\left(\frac{n\pi}{H}(z + H/2) \right), \tag{1.11}$$

where N is a normalization constant. The functions $f_m(\rho)$ are thus the solutions of

$$-\frac{\hbar^2}{2m^*}\left[\frac{1}{\rho}\frac{\partial}{\partial\rho}\left(\rho\frac{\partial f_m}{\partial\rho}\right) - \frac{m^2}{\rho^2}f_m\right] = \left(\varepsilon - \frac{\hbar^2 n^2 \pi^2}{2m^* H^2}\right)f_m$$

$$\rho^2 f_m'' + \rho f_m' + \left(\frac{\left(\varepsilon - \frac{\hbar^2 n^2 \pi^2}{2m^* H^2}\right)}{\frac{\hbar^2}{2m^*}}\rho^2 - m^2\right)f_m = 0. \qquad (1.12)$$

With the changes of variable and function

$$x = \rho\sqrt{\frac{2m^*}{\hbar^2}\left(\varepsilon - \frac{\hbar^2 n^2 \pi^2}{2m^* H^2}\right)}; \quad y_m(x) = f_m(\rho), \qquad (1.13)$$

we find that y_m fulfils the Bessel equation

$$x^2 y_m'' + x y_m' + (x^2 - m^2)y_m = 0. \qquad (1.14)$$

We retain the solutions that are regular at $x = 0$. They are $J_m(x)$, the Bessel functions of the first kind, for which there is $J_{-m}(x) = (-1)^m J_m(x)$. $J_{2p}(x)$ is even in x while $J_{2p+1}(x)$ is odd in x. We now write that the $f_m(\rho)$ functions vanish at $\rho = R_{box}$,

$$J_m\left(R_{box}\sqrt{\frac{2m^*}{\hbar^2}\left(\varepsilon_{nml} - \frac{\hbar^2 n^2 \pi^2}{2m^* H^2}\right)}\right) = 0, \quad \varepsilon_{nml} = \frac{\hbar^2 n^2 \pi^2}{2m^* H^2} + \frac{\hbar^2 \lambda_{mi}^2}{2m^* R_{box}^2}, \qquad (1.15)$$

where λ_{mi} is the ith zero of J_m: the λ_{mi} values are tabulated or can be easily computed, and are thus considered as known. Finally, a basis eigenstate is $|n, m, i\rangle$, where n labels the vertical motion, and m and i the radial ones. Let us note $\varepsilon_{n,m,i}$ are the corresponding large box energies.

Once we have the functions $\psi_{n,m,i}(\vec{r})$, the matrix elements of $V_{QD}(\vec{r})$ are calculated in this basis (the matrix elements of $H_{QD+box} - V_{QD}(\vec{r})$ are (by construction) diagonal in the $|n, m, i\rangle$ basis and equal to $\varepsilon_{n,m,i}$). Because the potential energy is rotationally invariant around the Oz axis, we know that the only non-vanishing elements of $V_{QD}(\vec{r})$ are between basis functions with the same m. They comprise two terms, which arise from the truncated cone and from the WL. The second one is a constant and so vanishes between states with different i values. Evaluating the matrix elements of $V_{QD}(\vec{r})$ is a lengthy process but straightforward; we shall not present this here explicitly.

In the following, we shall compare the outputs of the three methods in the specific case of a (Ga,In)As QD. The material parameters taken in the calculations are $m_{dot} = m_{barrier} = m^* = 0.067 m_0$ (m_0 = electron mass in vacuum) and $V_b = 413$ meV. The shape of the dot is a truncated cone with height $h = 3$ nm, up-radius $R_{up} = 8$ nm and basis radius $R_{down} = R = 10$ nm. In addition, this dot floats on a (Ga,In)As WL of thickness $d = 1$ nm. We find for the energy of the ground QD state in the three

Table 1.1. Comparison of the energy of the ground orbital states (1S), obtained by three different numerical methods.

Numerical method	$E_{1S}-E_{WL}$ in meV
Matrix diagonalization (12 000 basis states)	−250
Separable 'hybrid' method	−247
Simple variational estimate	−245

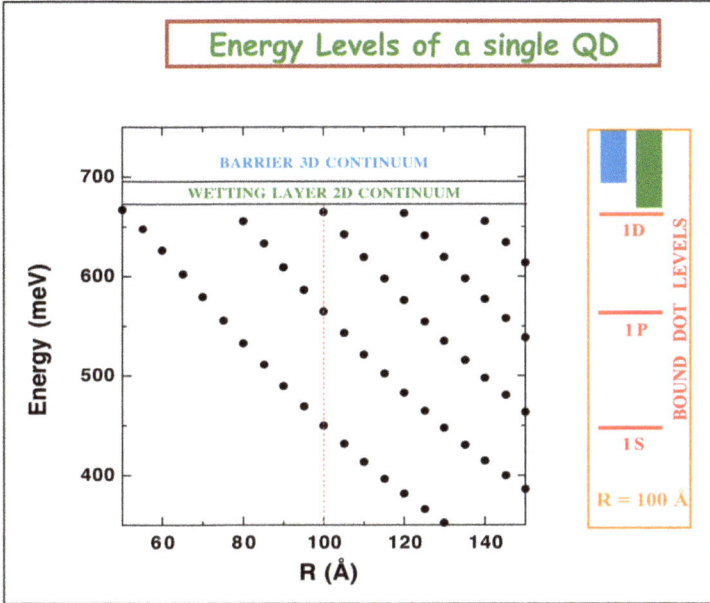

Figure 1.4. Calculated energies for states 1S, 1P, 1D,... for InAs cones with basis angle 12° (= $\pi/2-\alpha$); $V_b = 697$ meV; $m^* = 0.07 \, m_0$. On the right, levels scheme for the $R = 10$ nm dot, and energy positions of the WL and barrier continua (thick bars). Note that the energy origin is taken at the InAs conduction band edge.

descriptions the results presented in table 1.1, where E_{WL} is the ground energy of the thin quantum well (the WL) in the absence of a QD. As expected, the more elaborate method with its huge Hamiltonian matrix gives the best results (it is essentially exact). But the two other methods, much less numerically demanding, do rather well (2% off only). In addition, they have a big advantage in that they are physically transparent.

1.2.5 Examples of variational energies and wavefunctions

We show in figure 1.4 the calculated energy levels of a single QD obtained by the variational estimate (note that the energy origin is at the InAs conduction band edge). The dot geometry is an untruncated cone with basis angle 12°, and the figure presents variations with dot radius R of only the first bound state of each symmetry:

1S, 1P, 1D,... (big dots may also have 2S,... states). For these states one uses the three-parameter trial wavefunction

$$\psi_{1m}(\vec{r}) = \frac{e^{im\varphi}\rho^{|m|}}{\beta_{1m}^{1+|m|}\sqrt{\pi|m|!}\,\xi_{1m}\sqrt{\pi}} \exp\left\{-\frac{\rho^2}{2\beta_{1m}^2} - \frac{(z-z_{1m})^2}{2\xi_{1m}^2}\right\}, \qquad (1.16)$$

which generalizes the one in equation (1.3) for the 1S state. Note that the variational parameters depend only upon $|m|$.

We see that the number of bound states with different symmetries increases with increasing R. Also, for a given dot radius, states with increasing $|m|$ have increasing energies. The continuous energy bands above $\approx V_b$–27 meV and $\approx V_b$ correspond to unbound states in the WL (2D continuum, with a quantum-well confinement energy of $\approx$27 meV with respect to the GaAs barrier) and in the GaAs barrier (3D continuum, starting at the barrier energy edge), respectively. Note also that no bound state is found for QDs that are too small. Indeed, it is well-known in quantum mechanics that a short-range attractive 3D potential may not necessarily possess bound states. The existence of one (or more) bound level(s) relies not only on the characteristic size, but also on the strength of the confining potential and the effective mass of the carrier. Nanometric dots bind a few states mainly thanks to their important confining potentials (both in conduction and valence bands).

To give an insight into the spatial extension of the localized states, figure 1.5 presents the variational parameters for the 1S and 1P states in figure 1.4. For the untruncated cones, the dot height increases linearly with the radius: the wavefunctions

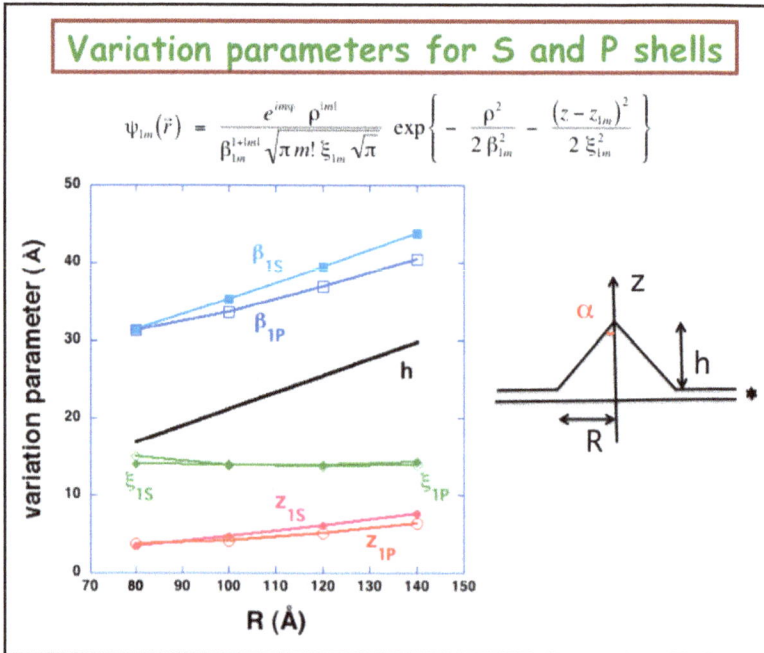

Figure 1.5. Variational parameters for states 1S and 1P of the InAs cones in figure 1.4.

expand radially and slightly displace upwards, while their vertical extensions remain almost unchanged in the interval of parameters considered.

The variational parameters depend strongly upon the carrier effective mass and confining potential energy and shape. For instance, for a (Ga,In)As/GaAs QD with $m^* = 0.045m_0$, $V_b = 261$ meV, $R_{down} = 16.5$ nm and $R_{up} = 12.3$ nm (thus an average radius 14.4 nm) we find for the 1S state: $\beta_{1S} = 7.1$ nm, $z_{1S} = 2.3$ nm and $\xi_{1S} = 2.75$ nm. The wavefunction is considerably more extended and pushed into the dot region as a consequence of the much lower energy barrier. The sizeable dependences upon the actual confining parameters simply reflect the strong quantum confinement regime prevailing in such nanometric systems.

Figure 1.6 shows iso-probability curves (i.e. curves for $|\psi(\rho, z)^2| = $ constant) of the three lowest bound states for an untruncated cone with $R = 10$ nm and a basis angle of $12°$ (the three levels 1S, 1P and 1D on the right-hand side of figure 1.4). The 1S state (lower panel) is very much localized in the cone and is maximum at $\rho = 0$, the

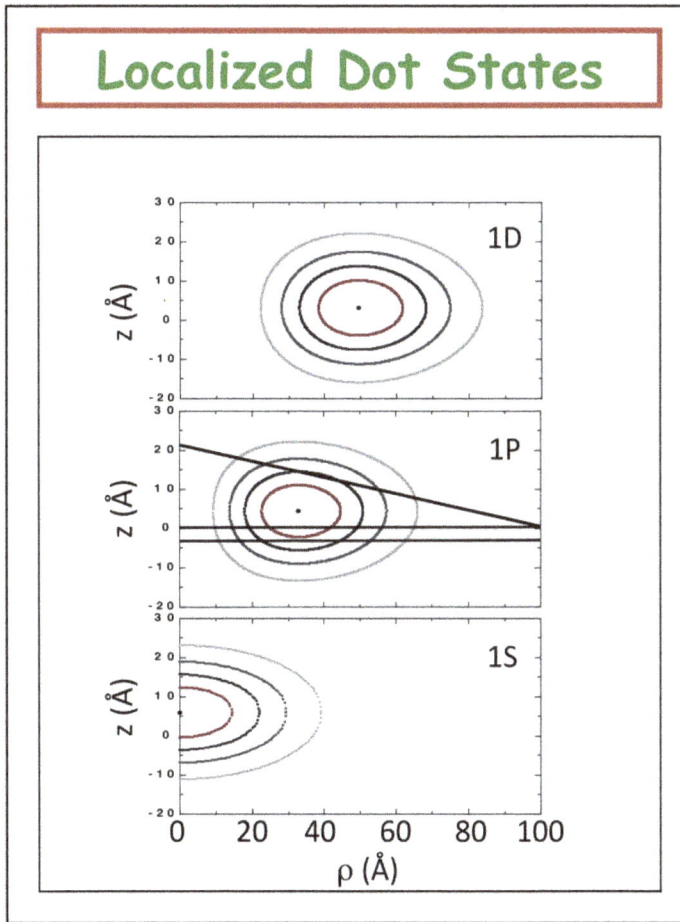

Figure 1.6. Iso-probability curves for the 1S (lower panel), 1P (middle panel) and 1D (upper panel) bound states on the right-hand side of figure 1.4. The limits of the QD and WL regions are shown in the middle panel.

centre of the cone in the xy-plane. For $m \neq 0$ the iso-probability vanishes for $\rho = 0$ and has a maximum that increases with $|m|$ because of the term $\rho^{|m|}$ in the trial wavefunction (equation (1.16)). We note in all cases the important penetration of the wavefunctions in the WL region as a consequence of the very small dot height. The centre of the wavefunction along Oz is roughly the same for the three states and slightly above the QD basis. In this model, all the wavefunctions have essentially the same vertical shape, while their radial part is driven further towards the exterior region for excited states: because of the strong aspect ratio, excited orbits 'escape' the QD by its lateral sides. The integrated probability of finding the electron in the dot decreases with increasing energy of the state: the S state is more bound to the dot than the P state. The D state is marginally bound and will be the more sensitive to what happens in the neighbourhood of the QD (traps,...).

1.3 Continuum states

Sustained investigations undertaken on QDs owe a lot to the existence of low-lying bound states that are well localized in the dots. However, the continuum states play a decisive part in many situations. For instance, the continuum states are the final states of the ionization transition where, by absorbing a photon, an electron initially in a state bound to the dot is sent into the continuum. Symmetrically, these are the initial states for the capture processes that load the QD with electrons and holes. These capture/ionization processes play a key part in devices that use QDs, such as lasers or photodetectors. They are also present in the 'crossed' optical interband transitions in the QD where one bound state for one band (say the valence band) is optically connected with the extended states of the other band (the conduction band). More generally, any particle or energy exchange between the QD and its environment depends sensitively on the characteristics of the continuum of states.

However, the description of QD extended states proves to be difficult for two main reasons. The first is that the QD potential is a very strong perturbation of the barrier matrix (since it creates bound states for electrons and holes: as mentioned previously, a short-range attractive 3D potential may not necessarily possess bound states). Thus, one may also expect that the continuum of states related to the 2D WL and to the 3D barriers underneath and above the dot, in particular those at low energy, will be much affected. In particular, one could expect that there will exist resonant states in the QD's continuum, because the QD potential shape has sharp edges (remember that a sharp 1D square potential well creates strong resonances—virtual bound states—in the continuum any time $kL = p\pi$, where L is the square well thickness and k the electron wavevector in the quantum well region). The probability density of such resonant states would pile up in the QD region, as compared to the other continuum states, and should thus play a prevalent role in the energy exchanges between the QD and its neighbourhood. The numerical implementation of codes for calculating the QD continuum states is often difficult on account of them being strongly attractive and short ranged, and also due to the very large number of plane waves needed to describe a small object with sharp edges. The second reason it is hard to compute the QD resonant states comes from the QD ensemble. In fact, the very strong localization of the QD low-lying states allows all the neighbouring and remote QDs to be neglected in their computation. This strong localization is absent for the continuum states, so their calculation may require many QDs to be accounted for at the same time. The lack of information on the size and parameters of the QDs make these calculations even more complex.

Approximations are therefore necessary in the evaluation (to be discussed later in this book) of the capture and relaxation of carriers by QDs. The most severe approximation consists in neglecting the QD and retaining the unperturbed WL and barrier states. In some cases, however, the existence of dot bound states can be accounted for by forcing the continuum states to be orthogonal to them. In this latter case we write (for the WL-related continuum)

$$\psi_{\vec{k}}(\vec{\rho}, z) = \frac{e^{i\vec{k}.\vec{\rho}}}{\sqrt{S}}\chi_{WL}(z)$$

$$\tilde{\psi}_{\vec{k}}(\vec{\rho}, z) \approx \psi_{\vec{k}}(\vec{\rho}, z) - \sum_n \langle\Phi_n(\vec{\rho}, z)||\psi_{\vec{k}}(\vec{\rho}, z)\rangle\Phi_n(\vec{\rho}, z), \qquad (1.17)$$

where the $\Phi_n(\vec{\rho}, z)$ are the bound states of the QD calculated by, say, the variational method outlined previously. At this level of approximation, the energy of the continuum states is not modified.

1.4 Coulombic interactions

Coulombic interactions play a special role in QDs. First, indiscernability effects come into play for identical carriers. Second, the coupling strength varies considerably depending on the two-carrier configurations: two bound, one bound and one delocalized and both free. Third, the energy spectra related to the first two of the three previous two-carriers configurations may overlap, leading to the appearance of resonances for both electrons (like for the He atom) and one electron–hole pair (*sui generis* to the condensed matter nature of the QD). Finally, a dot may host many electrons and/or holes, in neutral or charged states, which may to some extent be experimentally produced on demand. In the following we restrict ourselves to only those aspects of Coulombic states in a QD that will be relevant for the understanding and description of the capture and energy relaxation phenomena to be discussed in the next chapters.

1.4.1 Two-electron states

Considering the orbital and spin degeneracies, the 1S, 1P and 1D shells can accommodate altogether 10 electrons. However, storing several electrons in such a tiny volume leads to very significant Coulomb interactions between the carriers. The order of magnitude of this coupling can be roughly anticipated by considering two charges (of same sign or not) separated by an average distance that one can take as the variance,

$$\langle d_m \rangle \equiv \langle\psi_{1m}| \rho^2 + (z - z_0)^2 |\psi_{1m}\rangle^{1/2} = \left((1 + |m|!)\beta^2 + \xi^2/2\right)^{1/2}, \qquad (1.18)$$

where β and ξ are the radial and vertical extensions of the variational wavefunctions. One obtains $e^2/(4\pi\varepsilon_0\varepsilon_r\langle d_m\rangle) \approx 35, 31.5$ and 25.7 meV $(25.3, 23.7$ and 19.9 meV) for the 1S (1P) level at $R = 80, 100$ and 140 Å, respectively, using $\varepsilon_r = 12.5$ and the

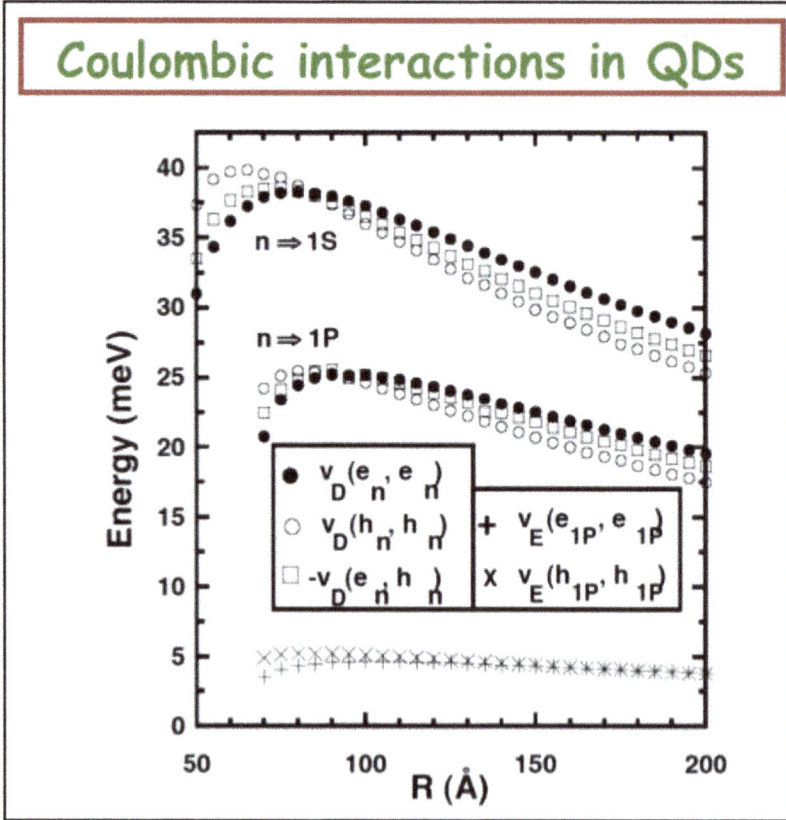

Figure 1.7. Coulomb integrals (direct v_D and exchange v_E) between two particles in a conical QD within the 1S or 1P shells, with the same QD parameters as figure 1.4. $\varepsilon_r = 12.5$.

1S- and 1P-related parameters β and ξ from figure 1.5. Figure 1.7 displays the calculated variations with the dot radius of the Coulomb energy between two charges in the same conical dot as in figures 1.4–1.6). Two electrons or two holes or one electron–hole pair have been considered.

The simple ansatz $e^2/(4\pi\varepsilon_0\varepsilon_r\langle d_m\rangle)$ provides a good first estimate of the Coulomb coupling values for the 1S and 1P shells, and also for their decreasing with increasing R. It clearly illustrates the main physical effect—namely, that the important Coulomb coupling in a dot is essentially related to the strong confinement of the interacting particles within the nanometric structure.

We note additionally that the direct interactions (v_D) for the different configurations are large (a few tens of meV), but they do not vary very much from one shell to the other, at least as compared to the one particle energy spacing between S and P shells. Also, the direct interactions between carriers are several times larger than the exchange energy v_E (between indiscernible particles). Indiscernibility effects are nevertheless important in quantum dots, and are discussed in the following.

We show in figure 1.8 the evolution of two-electron states in InAs-like QDs. The left-hand side column shows the non-interacting two-electron states built out of the

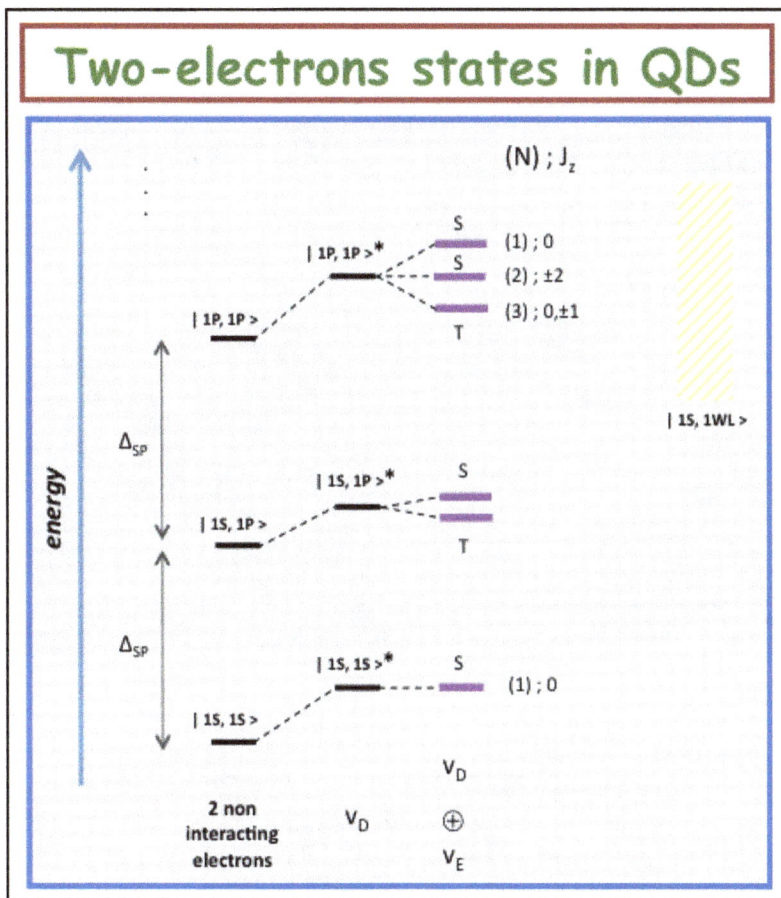

Figure 1.8. Schematic representation of the formation of two-electron states in a lens-shaped QD. The energy variations due to v_D and $v_D \oplus v_E$ are not at scale. See text.

one-electron states 1S and 1P when disregarding spin ($\Delta_{SP} = E_{1P} - E_{1S}$). Taking into account the direct term (v_D) of Coulomb correlations leads to the central column where a blue shift of all the states is found. Owing to the results in figure 1.7, this shift is roughly the same for all left-column states in figure 1.8. Taking into account the smaller exchange terms v_E gives rise to spin singlet (S) and spin triplet (T) states, corresponding to a total spin equal to 0 (singlet) and 1 (triplet), respectively. The degeneracy (N) and projection J_z of the total angular momentum (spin included) are also indicated in figure 1.8. Since the energy differences between the Coulombic corrections are smaller than the one-particle energy spacing Δ_{SP}, the two-electron final states display a pronounced 'shell' tag: SS, SP or PP. The weakness of the 'configuration mixing' will help us later to analyze the effects associated with two-electron states in QDs.

Figure 1.9 displays with more detail the build-up of two-electron states in the P shell. J and K refer to the direct and exchange integrals, respectively.

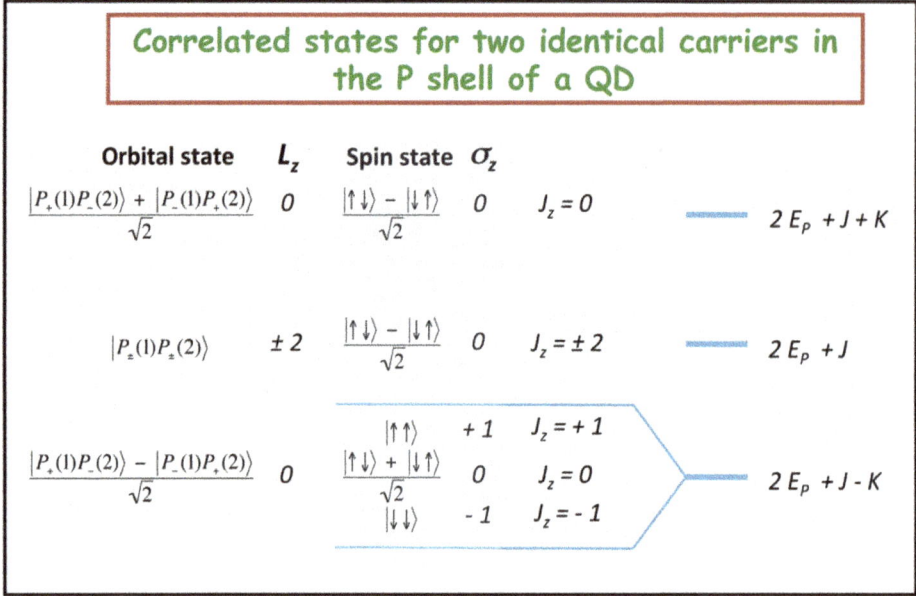

Figure 1.9. Build-up of two-electron states in the P shell of a lens-shaped InAs QD.

Using the orbitals $P_{\pm 1}$ one constructs three symmetric (P_+P_+, P_-P_- and ($P_+P_- + P_-P_+)/\sqrt{2}$) and one anti-symmetric (($P_+P_-—P_-P_+)/\sqrt{2}$) combinations, and the same for the two spins: three symmetric ($\uparrow\uparrow$, $\downarrow\downarrow$ and ($\uparrow\downarrow + \downarrow\uparrow)/\sqrt{2}$) and one anti-symmetric (($\uparrow\downarrow-\downarrow\uparrow)/\sqrt{2}$) combinations. That makes in all six two-electron states, anti-symmetric regarding both orbital and spin labels. Let us consider the Coulomb couplings among these states. Firstly, since Coulomb interaction V_c (1, 2) is spin-independent, the spin component is conserved. Moreover, since it also displays cylindrical orbital symmetry, the total L_z component conserves. As a consequence of these two properties, there are only diagonal matrix elements, and we obtain the two sets of results

$$\Psi_{n,m,\lambda} = \frac{1}{\sqrt{2}}\{P_n(1)P_m(2) + \lambda P_n(2)P_m(1)\}, \quad \lambda = \pm 1$$

$$\Rightarrow \langle \Psi_{n,m,\lambda} |V_c|\Psi_{n,m,\lambda}\rangle = J_1 + \lambda K \tag{1.19a}$$

$$\begin{cases} J_1 = \langle P_n(1)P_m(2)|V_c(1, 2)|P_n(1)P_m(2)\rangle \\ K = \langle P_n(1)P_m(2)|V_c(1, 2)|P_n(2)P_m(1)\rangle, \end{cases}$$

for $n \neq m$, and

$$\Psi_{n,n,-} = P_n(1)P_n(2)$$

$$\Rightarrow \langle \Psi_{n,n,-} |V_c|\Psi_{n,n,-}\rangle = J_2 = \langle P_n(1)P_n(2)|V_c(1, 2)|P_n(1)P_n(2)\rangle, \tag{1.19b}$$

for $n = m$, where $n, m = \pm 1$ for the P orbitals and $\lambda = +1$ for a symmetric (-1 for an anti-symmetric) two-electron orbital wavefunction. Finally, one can easily show that $J_1 = J_2$, which explains the final classification shown in figure 1.9.

It is interesting to point out that the low-lying states of the two-electron continuum are not built out of unbound single particle states. Rather, they are formed by letting one electron be unbound while the second is in the ground bound state of the QD. These are the states labelled $|1S;WL\rangle$ in figure 1.8. This feature has very important consequences on the QD physics. In particular, the two-electron states built out of single-particle 1P states can (depending on the QD parameters (size,...)) lie inside the $|1S;WL\rangle$ continuum. This situation is reminiscent of the excited states found in atomic He.

1.4.2 Electron–hole pairs

We will also be interested in the interband states formed by one interacting electron–hole pair. As we shall discuss later, their excited states play an important part in the physics of capture and/or relaxation of carriers by the QDs. The eigenstates of an electron–hole pair are sketched in figure 1.10. They are classified into bound states, extended states and 'crossed' states:

- *Bound states.* The bound states appear on the low-energy side of the energy spectrum. They most clearly exemplify the simultaneous confinements of the two carriers by the QD. The ground interband transition (full-red features in figure 1.10) always belongs to the bound states species (in InAs QDs).

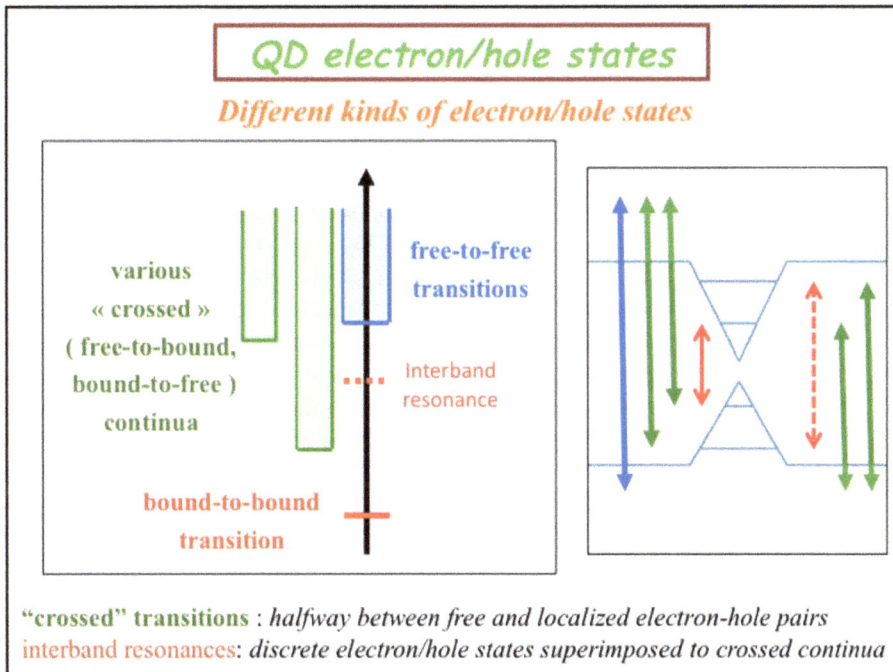

Figure 1.10. Sketch of the three kinds of electron–hole pair states in a QD.

- *High-energy continuum.* The high-energy extended pair states (blue features in figure 1.10) are those of the dissociated pairs either in the 2D WL states or in the 3D GaAs-like states; at high energy they are little affected by the QD but they display multiple scattering effects when their kinetic energy (as measured from the dissociation edge) is much smaller than the sum of the barrier heights.
- *'Crossed' states.* Finally, the 'crossed' states also correspond to dissociated pairs since in a 'crossed' state one of the particles belongs to the discrete spectrum while the other is in an extended state. Like in the two-electron states discussed previously, the 'crossed' electron–hole states play an important part in the QD embedding in its environment. They also play a unique part in capture and relaxation effects (to be discussed in the next chapters). They create resonances in the interband spectrum of a QD. The resonances show up when an electron–hole bound state happens to fall in the continuum of 'crossed' electron–hole states: this is illustrated in figure 1.10, where we observe the presence of green crossed transitions at the same energy as the dashed red line involving only bound states.

1.4.3 Multi-charged dots

Let us finally stress the fact that a QD can store many electron–hole complexes, either neutral, like poly electron–hole pairs, or charged with N electrons and M holes (see, for example, Lelong and Bastard (1996)). The storing of poly electron–hole pairs comes from the fact that attractive electron–hole interactions nearly balance repulsive electron–electron or hole–hole interactions (see figure 1.7). In the case of two correlated electron–hole pairs (a 'bi-exciton') in the ground state there are four attractive interactions (e1 with h1 and h2; e2 with h1 and h2) while there are two repulsive interactions (e1–e2 and h1–h2). Similarly, for the complex with two electrons and one hole, all in the respective S state) there are two attractive (h with e1 and e2) and one repulsive interactions. When this complex recombines, one possible channel consists in leaving the extra electron in the ground state. The energy position of this recombination line, compared to that of a single electron–hole pair, depends sensitively on the QD shape and parameters. Similarly, it is known that the complex with two electrons and two holes can radiatively decay while leaving a single electron–hole pair bound to the QD. The bi-pair line can be red or blue shifted with respect to the recombination line of a single electron–hole pair. It is important to stress that it is possible to have an 'unbound' bi-pair (i.e. one that fluoresces at higher energy than the single pair), that does not bring any extra difficulty compared to the case where it is 'bound' (fluoresces at lower energy than the single pair). This contrasts sharply with bulk or 2D heterostructures where due to the continuum of absorption such an unbound complex would immediately dissociate and would not be observable. We shall however not consider multi-charged dots further in the remaining chapters of this book.

1.5 Phonons and polarons in InAs quantum dots

In the following we present a detailed discussion of the particular coupling between charged carriers (electrons and holes) with the longitudinal optical (LO) phonons of

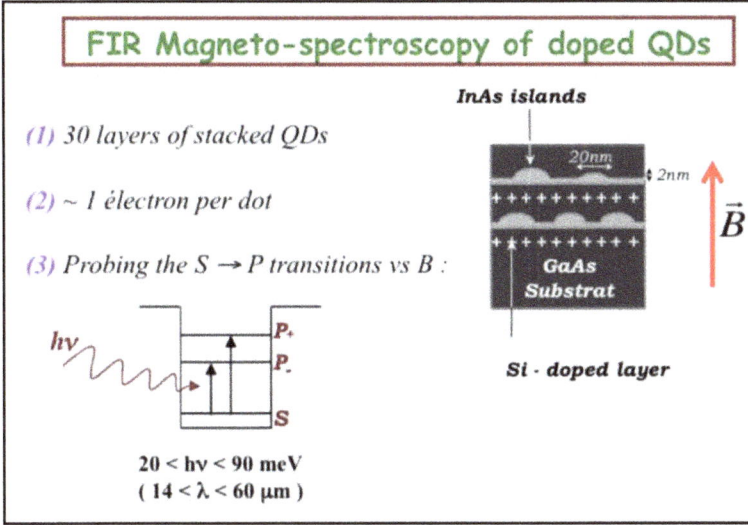

Figure 1.11. Scheme of magneto-transmission experiments in doped QDs.

the QDs and of the crystalline matrix where the QDs are embedded. We shall show that the true elementary excitations of the self-assembled QDs correspond to a coherent mixture of electron and LO phonon states (polarons) that results from their strong coupling, instead of the usual unperturbed electron states weakly damped by electron–LO phonon interaction handled by the Fermi golden rule. The reason to insist in this book on the strong coupling regime lies, as we shall discuss at length below, in the irreversible evolution brought about by the genuine polaron instability associated with vibration anharmonicity. This particular route to energy relaxation is, to our knowledge, completely novel in III–V heterolayers. It actually ensures a very efficient capture and, more importantly, solves the puzzle of one-electron energy relaxation through the excited bound states of a self-assembled QD down to its ground level. In this chapter, we introduce and develop the main notions and features of the stationary quantum-dot-polaron eigenstates that will serve to describe in chapter 3 the relaxation mechanism within the polaron formalism.

The unusual coupling between electrons and LO phonons was shown by Hameau *et al* (1999) by means of magneto-transmission experiments. Figure 1.11 summarizes the experiment: it consists of measuring the transmission of far-infrared (FIR) light that propagates parallel to the growth axis and interacts with ≈ 30 layers of QDs that have been modulation-doped by donors, with a doping content adjusted so as to provide only one electron per QD. At thermal equilibrium the electrons loaded in the QD occupy the dot ground state 1S at low enough temperature. The light probes the empty excited states of the QD. Since the photon transfers $\pm\hbar$ to the electron, the transitions are between the S and P shells: $|1S\rangle \rightarrow |1P_{\pm}\rangle$ (since the $1P_{\pm}$ states fulfil $L_z|P_{\pm}\rangle = \pm\hbar|1P_{\pm}\rangle$), like for a 2D H atom. Here, we shall not consider the absorption spectrum of the dot ensemble in detail, but only be interested in the energy position of the peak of the broadened absorption line. For a rotationally invariant dot, the two transitions $|1S\rangle \rightarrow |1P_{\pm}\rangle$ occur at the same energy, i.e. at $\hbar\omega_{\pm} = \Delta_{SP} = E_P - E_S$.

An important specificity of this experiment is that it allows light transmission to be varied by applying a strong magnetic field parallel to the growth axis (applying it in the layer plane is not so interesting). Before presenting the results of measurements, let us discuss the modelling of the magneto-absorption.

1.5.1 Modelling the far-infrared absorption: the 'artificial atom' picture

To study the field effect on the QD states we introduce the vector potential $\vec{A}(\vec{r})$ such that $\vec{B} = \vec{\nabla} \times \vec{A}$ and choose, like in atomic physics, the symmetrical gauge $\vec{A}(\vec{r}) = \vec{B} x \vec{r}/2$. We replace $\vec{p}$ by $\vec{p} + e\vec{A}(\vec{r})$ (with $e > 0$ the positron charge) in the effective QD Hamiltonian and finally obtain

$$H_{QD}(\vec{B}) = H_{QD}(0) + H_{par} + H_{dia}, \tag{1.20a}$$

$$H_{par} = \frac{\omega_c L_z}{2}, \qquad H_{dia} = \frac{m^* \omega_c^2}{8}(x^2 + y^2), \qquad \omega_c = \frac{eB}{m^*}, \tag{1.20b}$$

where $H_{QD}(0) \equiv H_{QD}$ is given in equation (1.1), L_z in equation (1.2) and ω_c is the cyclotron frequency. One recognizes the linear paramagnetic (H_{par}) and quadratic diamagnetic (H_{dia}) Zeeman shifts. The former is independent of the QD parameters whereas the latter is very sensitive to the in-plane extension of the QD wavefunction. At the lowest order in perturbation one has

$$\left(H_{par}\right)_{1m} = \left\langle \psi_{1m} | H_{par} | \psi_{1m} \right\rangle = m\hbar\omega_c/2, \tag{1.21a}$$

$$(H_{dia})_{1m} = \left\langle \psi_{1m} | H_{dia} | \psi_{1m} \right\rangle \approx m^* \omega_c^2 \beta_{1m}^2 [1 + |m|!]/8, \tag{1.21b}$$

for the separable trial function ψ_{1m} (equation (1.16)). Table 1.2 gives the values of the linear and quadratic shifts at high magnetic field $B = 20$ T for an $R = 100$ Å dot (with variational parameters $\beta_{1S} \approx \beta_{1P} \approx 34$ Å; see figure 1.5).

Note that the diamagnetic shift remains small compared to the S–P energy distance $\Delta_{SP} \approx 50$ meV. The $1P_\pm$ states have an important linear variation with the field, much larger than the diamagnetic contribution and definitely not negligible as compared to Δ_{SP}. Treating the magnetic shifts in perturbation leads to the B-dependent transition energies

$$\hbar\omega_\pm(B) = \hbar\omega_\pm(0) \pm \frac{1}{2}\hbar\omega_c + \gamma B^2, \tag{1.22a}$$

Table 1.2. Linear and quadratic shifts ($H_{par})_{1m}$ and $(H_{dia})_{1m}$ for the 1S ($m = 0$) and $1P_\pm$ ($m = \pm 1$) states of a dot with $R = 100$ Å at $B = 20$ T.

	Linear (meV)	Quadratic (meV)
1S	0	1.5
$1P_\pm$	± 17	3.0

where $\hbar\omega_\pm(0) = \Delta_{SP}$ and $\gamma = \gamma_P - \gamma_S$ is the difference between the diamagnetic shifts of the P and S states,

$$\gamma_{1m} = \frac{e^2}{8m^*}\langle\psi_{1m}|x^2 + y^2|\psi_{1m}\rangle. \qquad (1.22b)$$

The quadratic term is actually small for nanometric dots even at high fields: $\gamma B^2 \approx$ few meV at $B = 20$ T. The B dependence of the inter-shell transition energy is accordingly essentially given by the linear paramagnetic contribution.

The previous model of optical transitions in a QD is frequently coined either the 'artificial atom' model because of the similarities between its predictions and those of the Zeeman-shifted atomic transitions in actual atoms, or 'macro-atom' model, because dots are considerably bigger than atoms. Life would indeed be simple if, by packing millions of atoms to make a semiconductor QD, one could recover all and uniquely the properties of a very big atom. Indeed, as we shall see in what follows, the underlying condensed matter nature of a QD system brings specific and crucial features to the dots' optical response.

1.5.2 Far-infrared absorption at zero field and anisotropic macro-atom model

Figure 1.13 shows the optical transmission spectrum measured at $B = 0$ for an electromagnetic (*e.m.*) wave propagating along the growth axis and linearly polarized along three different directions in the layer plane. Clearly, the [110] and [1–10] directions are not equivalent since the resonant energies (that correspond to transmission minima) are different. The anisotropic absorption is unexplainable with a Hamiltonian that is assumed to be rotationally invariant around the growth axis. Note however that the anisotropy is weak: the resonant frequencies differ by

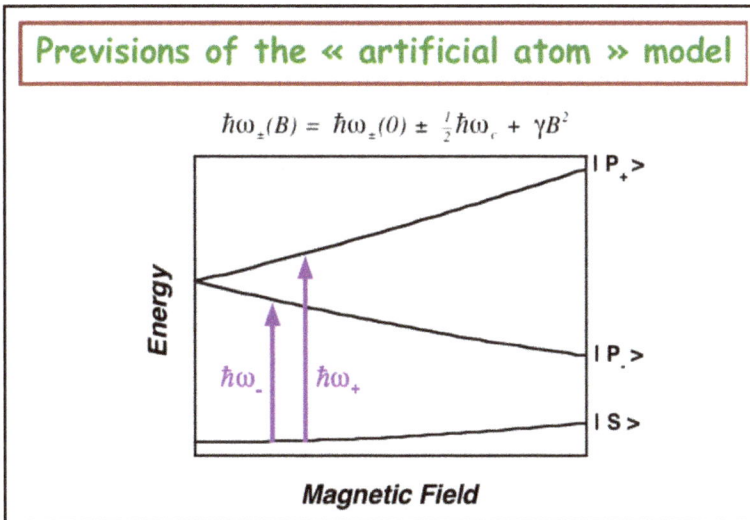

Figure 1.12. Sketch of the magnetic field dependences of the optical transition energies $\hbar\omega_\pm$ in a lens-shaped InAs QD, in the 'artificial atom' approximation (see text).

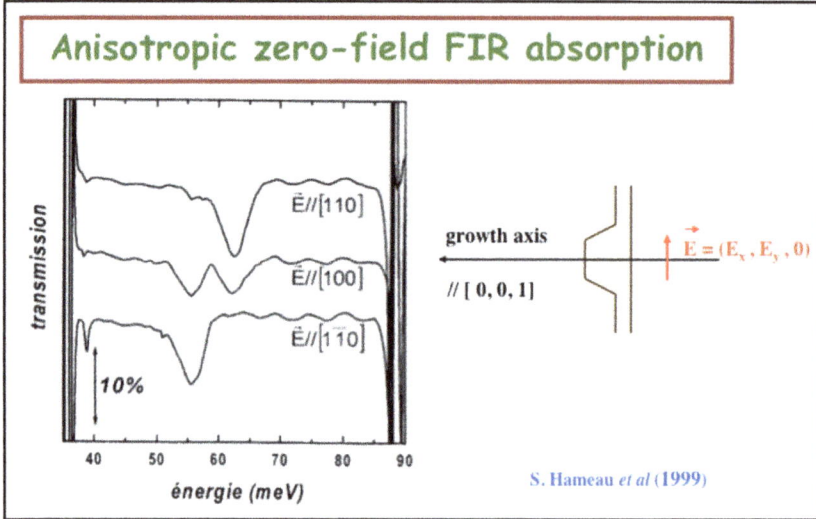

Figure 1.13. Zero magnetic field transmission of an ensemble of InAs self-assembled QDs versus photon energy for three different linear polarizations of the *e.m.* wave. $T = 4$ K. From Hameau *et al* (1999).

less than $\approx 15\%$. Several effects may cause a small departure from the assumed cylindrical symmetry:

- At the 'atomic' level, when one considers the atomic orbitals there is no cylindrical symmetry (the zinc-blende lattice leaves [110] and [1–10] equivalent only in a bulk and unstrained material: neither of these two conditions apply to InAs QDs).
- At the level of the envelope function, the potential energy may display anisotropy. For example, the truncated cones may have an elliptic basis instead of a circular one (see discussion in section 1.2.1).

Let us further consider the second effect. One can check that a lateral shape that is slightly elliptical leads to a non-vanishing $\langle \psi_{+1} | \delta V | \psi_{-1} \rangle$, where δV is the difference in the confining potential obtained by assuming truncated cones with elliptical or circular basis. On the other hand, there is no diagonal shift of the $P_{\pm 1}$ states' energy up to the linear order in ε, where ε is the relative departure of the dimensions with respect to their average (see scheme in figure 1.14). Hence, knowing that the S–P transition energy is large (50 meV or so), we can restrict the effect of δV to the 2×2 subspace generated by the $P_{\pm 1}$ states of the circular dot,

$$
\begin{vmatrix} E_P + \dfrac{1}{2}\hbar\omega_c + \gamma_P B^2 - E & \langle \psi_{+1} | \delta V | \psi_{-1} \rangle \\[2mm] \langle \psi_{-1} | \delta V | \psi_{+1} \rangle & E_P - \dfrac{1}{2}\hbar\omega_c + \gamma_P B^2 - E \end{vmatrix}. \tag{1.23a}
$$

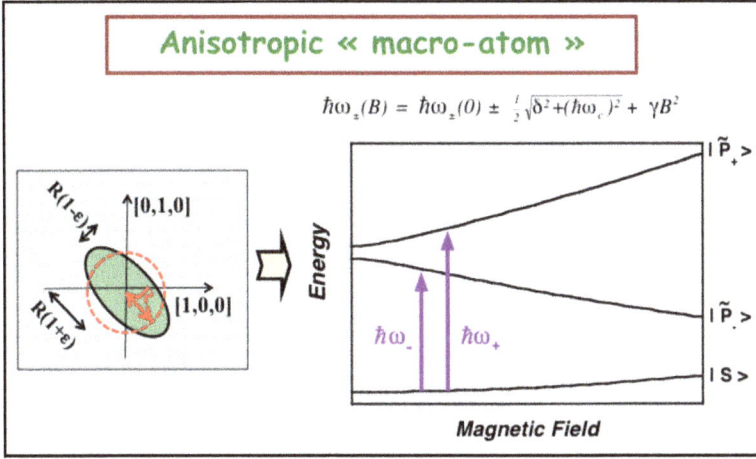

Figure 1.14. Left: scheme of an anisotropic dot (basis cross-section). Right: sketch of the magnetic field dependence of the transition energies in an anisotropic 'macro-atom'. The tilde refers to the new P states and the + (−) to the upper (lower) branch.

We find a lifting of the degeneracy and the energies of the two P states equal to

$$E_\pm = E_P + \gamma_P B^2 \pm \sqrt{(\hbar\omega_c)^2 + \delta^2}$$
$$\delta = 2\left|\langle\psi_{+1}\,|\,\delta V\,|\psi_{-1}\rangle\right|. \qquad (1.23b)$$

This results in transition energies that can be written (in lieu of equation (1.22)) as

$$\hbar\omega_\pm(B) = \hbar\omega_\pm(0) \pm \frac{1}{2}\sqrt{\delta^2 + (\hbar\omega_c)^2} + \gamma B^2. \qquad (1.24)$$

The magnetic field dependence of the expected transition energies in a laterally anisotropic 'macro-atom' is shown in figure 1.14. Like in the isotropic case, we find an 'ascending' and a 'descending' branch. However, the field dependence at low field ($\hbar\omega_c \ll \delta$) is quadratic in B instead of being linear if $\delta = 0$.

1.5.3 Far-infrared magneto-absorption and breakdown of the macro-atom model

Figure 1.15 shows the transmission spectra (left panel) obtained for different values of B between 9 T and 28 T for the same sample as used to create figure 1.13. There are two regions where the transmission is very small: near 36 meV, the *restrahlen* band of the bulk GaAs substrate (i.e. the light absorption by bulk *transverse* optical phonons), and near 90 meV due to parasitic absorption of the experimental setup. The transmission minima versus B are collected (fan-chart) on the right-hand side of figure 1.15. The dashed line corresponds to the best fit of the experimental data to the model of a laterally anisotropic 'artificial atom'. There is agreement on the existence of two branches. However, there are pronounced deviations between the data and the anisotropic 'artificial atom': the ascending branch displays pronounced

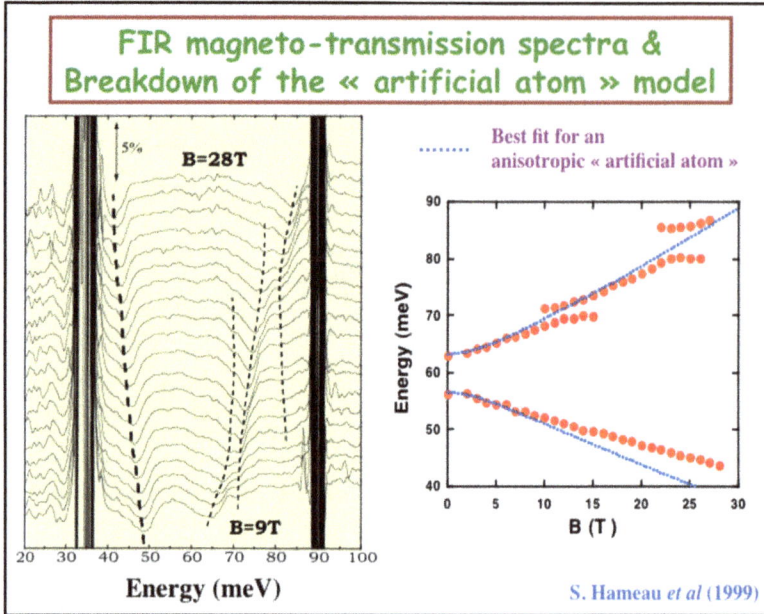

Figure 1.15. Magneto-trasmission spectra (left) and fan-chart (right) of the magneto-optical transitions in InAs QDs. $T = 4$ K. The dashed lines in the left panel are guides for the eyes. The dashed lines in the right panel are the best-fit using the anisotropic macro-atom model. Adapted from Hameau *et al* (1999).

anticrossings; and there is no mirror effect between the two branches since the descending branch has no anticrossings and a far too shallow slope.

1.5.4 Qualitative evidence for longitudinal optical phonons effects

A first hint of LO phonon participation in the FIR absorption of InAs QDs is shown in figure 1.16.

We find that the LO phonon energy $\hbar\omega_{LO} \approx 36$ meV (characteristic of bulk GaAs and not far from the one of strained InAs) recurrently shows up in the experimental results in relation to the deviations between modelling and experiments. For instance, the descending branch increasingly deviates from the expected behaviour when the photon energy approaches $\hbar\omega_{LO}$. Similarly, the first anticrossing on the ascending branch (near 12 T) shows up when the photon energy is near $2\hbar\omega_{LO}$. Finally, the larger anticrossing (near 24 T) shows up when the ascending and descending branches are split by $\hbar\omega_{LO}$. We therefore need to investigate more thoroughly the coupling between the electrons confined in a QD and the LO phonons. However, to set the stage more clearly, let us quickly recall a few properties of phonon dispersions in bulk semiconductors.

1.5.5 Phonons in bulk III–V materials: two essential features

III–V cubic compounds crystallize in the zinc-blende lattice. Figure 1.17 presents the phonon modes in GaAs. One observes two kinds of dispersion near the centre-zone region (at $\zeta = 0$ in the figure), which result from the existence of two different atoms

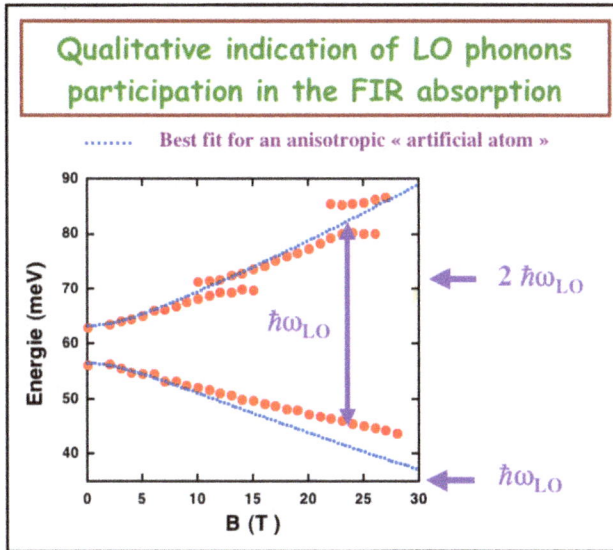

Figure 1.16. Symbols and dashed lines: fan-chart of magneto-optical transitions in InAs QDs and best fit to the anisotropic 'artificial atom' (same results as in the right panel of figure 1.15). Arrows: indications of the particular role played by the LO phonons in the magneto-transmission spectrum. Adapted from Hameau *et al* (1999).

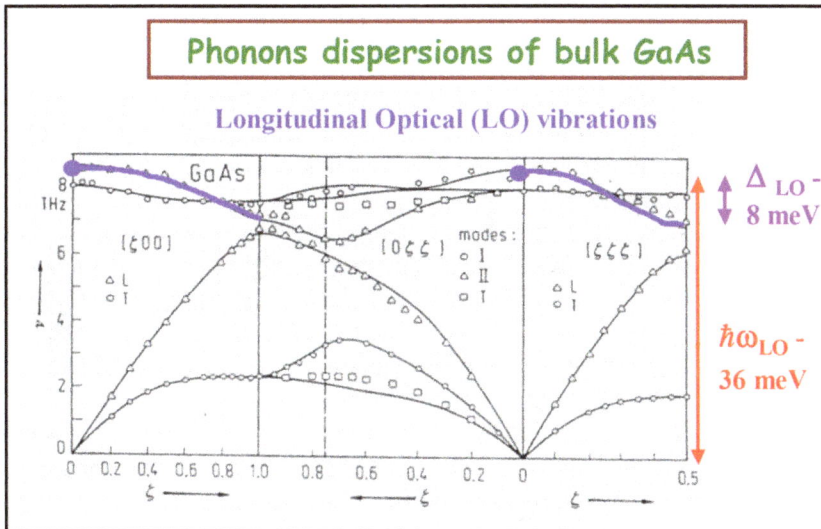

Figure 1.17. Phonon dispersions in GaAs along high symmetry directions. Adapted from Landolt–Börnstein (1970–2014).

per unit cell: acoustical (low-energy) and optical (high-energy) modes. These modes have distinct characteristics. For small wavevectors there is: a linear dispersion for the acoustical vibrations while the optical ones are almost dispersionless; and the two atoms in a lattice cell oscillate in phase for the low-energy acoustical vibrations but in phase opposition for the optical vibrations (see figure 1.18).

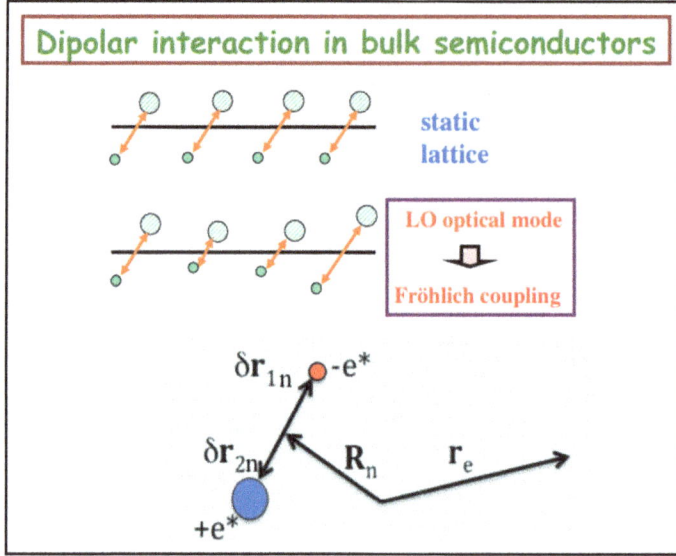

Figure 1.18. Schematic representation of atomic LO vibrations in a lattice with two different atoms per unit cell, and notation of the relative arrangements of the atoms in one arbitrary (the nth) unit cell of the 3D crystal.

1.5.6 Fröhlich coupling between electrons and longitudinal optical phonons in bulk

In the presence of lattice vibrations the crystal potential is no longer periodic, so the electron state is perturbed by the excited (quantized) lattice. The different kinds of atomic motions for the acoustical and optical modes lead to different electron–lattice couplings. In the following we focus on the so-called Fröhlich coupling between electrons and optical vibrations.

Since the two atoms in a unit cell are different, the bonds are not strictly covalent but slightly polar: in GaAs, the Ga (As) atoms have a slight positive e^* (negative $-e^*$) charge. Hence, when the two atoms vibrate in phase opposition, they create a long-range electric field that oscillates in time at the LO phonon frequency. The Fröhlich interaction results from the coupling of the electron with the total field from all unit cells (see, for example, Yu and Cardona (1999)).

Let us consider first a classical description of this interaction. For this purpose, we denote by $\vec{R}$, $\delta\vec{n}_n$ and $\delta\vec{r}_{2n}$ and $\vec{r}_e$ the positions of the nth unit cell, of atoms 1 and 2 in this cell and of a conduction electron, respectively (see figure 1.18). The contribution of the charges in the nth unit cell to the potential energy of the electron is

$$V_n\left(\vec{r}_e, \vec{R}_n\right) = -e\left(\frac{e^*}{4\pi\varepsilon_0\varepsilon_r\left|\vec{r}_e - \vec{R}_n - \delta\vec{r}_{2n}\right|} - \frac{e^*}{4\pi\varepsilon_0\varepsilon_r\left|\vec{r}_e - \vec{R}_n\right| - \delta\vec{r}_{1n}\right|}\right)$$

$$\approx K\delta\vec{r}_n \cdot \frac{\partial}{\partial\vec{r}_e}\frac{1}{\left|\vec{r}_e - \vec{R}_n\right|} = -K\frac{\delta\vec{r}_n \cdot \left(\vec{r}_e - \vec{R}_n\right)}{\left|\vec{r}_e - \vec{R}_n\right|^3}, \tag{1.25}$$

where $K = ee^*/[4\pi\varepsilon_0\varepsilon_r]$ and $\delta\vec{r}_n = \delta\vec{r}_{2n} - \delta\vec{r}_{1n}$. The long distance approximation $|\delta\vec{r}_{1n}|, |\delta\vec{r}_{2n}| \ll |\vec{r}_e - \vec{R}_n|$ has been made. This also allows the solid to be considered as a homogeneous dielectric characterized by a relative permittivity ε_r. Due to the atoms' motions, $\delta\vec{r}_n$ is time-dependent. More precisely, $\delta\vec{r}_n(t)$ has both a static part $\delta\vec{r}_n^{\text{sta}}$ (that contributes to the build-up of the stationary Bloch states and will no longer be considered) and a time-dependent contribution $\vec{\delta}_n(t)$ that induces transitions between the Bloch states.

Searching for a travelling wave for the relative displacement, we write

$$\vec{\delta}_n(t) = \frac{1}{2}\vec{\delta}_0 e^{i(\vec{q}\cdot\vec{R}_n - \omega t)} + \text{h. c.} \tag{1.26}$$

Then summing over $\vec{R}_n$, transforming the discrete sum into an integration, using $\frac{\partial F(\vec{r}_e - \vec{R}_n)}{\partial \vec{r}_e} = -\frac{\partial F(\vec{r}_e - \vec{R}_n)}{\partial \vec{R}_n}$ and integrating by parts, one has

$$V(\vec{r}_e) = \sum_n V_n(\vec{r}_e - \vec{R}_n, t) \rightarrow \int \frac{d^3}{\Omega_0} V_n(\vec{r}_e - \vec{R}_n, t) =$$

$$\approx K\frac{\vec{\delta}_0}{2} \cdot \int \frac{d^3 R_n}{\Omega_0}\left[\frac{\partial}{\partial\vec{R}_n}e^{i(\vec{q}\cdot\vec{R}_n - \omega t)}\right]\frac{1}{|\vec{r}_e - \vec{R}_n|} + hc$$

$$= \frac{i2\pi K}{\Omega_0}\frac{\vec{\delta}_0 \cdot \vec{q}}{q^2}e^{i(\vec{q}\cdot\vec{r}_e - \omega t)} + hc, \tag{1.27}$$

where Ω_0 is the unit cell volume and we have used the 3D Fourier transform $\int d^3r\frac{e^{-i\vec{k}\cdot\vec{r}}}{|\vec{r}|} = 4\pi/k^2$. The final expression displays the two salient features of the Fröhlich coupling: only the longitudinal phonon modes are coupled to the electron; and the interaction is long range (its Fourier transform varies like $1/q$).

The final shape of the Fröhlich interaction is obtained after quantizing the phonon field. If we denote by $\hbar\omega_\alpha(\vec{q})$ the energy of the phonon with polarization α and wavevector $\vec{q}$, quantization of the vibration modes leads to defining creation $(a_{\alpha\vec{q}}^+)$ and annihilation $(\alpha_{\alpha\vec{q}})$ operators for phonons. They respectively increase (decrease) the number of phonons in the $(\alpha, \vec{q})$ mode by one. The operator number in the $(\alpha, \vec{q})$ mode $n_{\alpha,\vec{q}} = a_{\alpha\vec{q}}^+a_{\alpha\vec{q}}$ and there is

$$a_{\alpha\vec{q}}^+\left|n_{\alpha\vec{q}}\right\rangle = \sqrt{n_{\alpha\vec{q}} + 1}\left|n_{\alpha\vec{q}} + 1\right\rangle; \quad a_{\alpha\vec{q}}\left|n_{\alpha\vec{q}}\right\rangle = \sqrt{n_{\alpha\vec{q}}}\left|n_{\alpha\vec{q}} - 1\right\rangle$$

$$H_{\text{phonon}} = \sum_{\alpha,\vec{q}}\hbar\omega_{\alpha\vec{q}}\left(n_{\alpha\vec{q}} + \frac{1}{2}\right). \tag{1.28}$$

The Fröhlich Hamiltonian can be written (omitting the $\alpha = \text{LO}$ label) as

$$H_{\text{e-ph}} = \sum_{\vec{Q}}\left\{V_{\vec{Q}}^*e^{-i\vec{Q}\cdot\vec{r}}a_{\vec{Q}}^+ + V_{\vec{Q}}e^{+i\vec{Q}\cdot\vec{r}}a_{\vec{Q}}\right\}, \tag{1.29a}$$

Table 1.3. Value of the Fröhlich constant for a few polar materials.

Compound	NaF	AgBr	ZnO	GaAs	InSb
α_F	6.3	1.6	0.85	0.06	0.014

$$V_{\vec{Q}} = -i\frac{C_F}{Q\sqrt{\Omega_{cr}}}, \quad C_F = e\sqrt{\frac{\hbar\omega_{LO}}{2\varepsilon_0}\left(\frac{1}{\varepsilon_r(\infty)} - \frac{1}{\varepsilon_r(0)}\right)}, \tag{1.29b}$$

where Ω_{cr} is the crystal volume, ω_{LO} the frequency of the LO modes and $\varepsilon_r(\infty)$ ($\varepsilon_r(0)$) the high-frequency (the static) dielectric permittivity. The strength of the electron–LO-phonon interaction has the same $1/Q$ dependence as in equation (1.27), and C_F can be related to the classical charge e^*. However, the importance of the Fröhlich coupling on electronic properties is usually measured by a different parameter than C_F—namely, the Fröhlich constant α_F defined as

$$\alpha_F = \frac{e^2}{4\pi\varepsilon_0\hbar}\sqrt{\frac{m^*}{2\hbar\omega_{LO}}}\left(\frac{1}{\varepsilon_r(\infty)} - \frac{1}{\varepsilon_r(0)}\right). \tag{1.30}$$

Table 1.3 gives α_F for a few materials. Moreover, for GaAs one has $\varepsilon_r(\infty) \approx 10.9$, $\varepsilon_r(0) \approx 12.5$ and thus $C_F = 196.1$ meV $\text{Å}^{1/2}$.

Because α_F is small in the III–V compounds we are interested in, the electron–LO phonon interaction is usually treated in perturbation (see, for example, Ridley (1988)). In this context, Fröhlich has provided the first weak-coupling ($\alpha_F \to 0$) perturbation theory results to the interacting electron–LO phonon system, by finding a shift equal to $-\alpha_F\hbar\omega_{LO}$ to the energy of low-energy electrons, thereby giving physical significance to the constant α_F.

Despite the small value of α_F, the Fröhlich interaction is responsible for the most efficient mechanism to relax the energy of high-energy electrons (i.e. with E > $\hbar\omega_{LO}$), with sub-picosecond scattering times measured in bulk GaAs as well as in GaAs quantum wells. Such important rates can nevertheless be well explained by using the perturbative Fermi golden rule.

In the next paragraph, we consider the case of QDs.

1.5.7 Phonons and Fröhlich interaction in a dot: a few assumptions

One may wonder about the pertinence of employing the previous bulk-related properties to describe the interaction between one electron and the lattice vibrations in a QD, as will be done below. Indeed, one would expect that the actual vibration modes of a dot would depend on its shape and actual composition and strain profiles, which, as previously mentioned, are unknown in actual samples. Instead, we shall retain the bulk phonon modes (i.e. we shall neglect the nano-structuration of the phonon modes by the QDs) and assume a bulk-like Fröhlich Hamiltonian since, as shown in figure 1.19, all the ingredients of a Fröhlich coupling are present in actual QDs. The only acknowledgement of the QD peculiarities will be to use an effective Fröhlich constant. As we shall see later on, the excellent agreement between

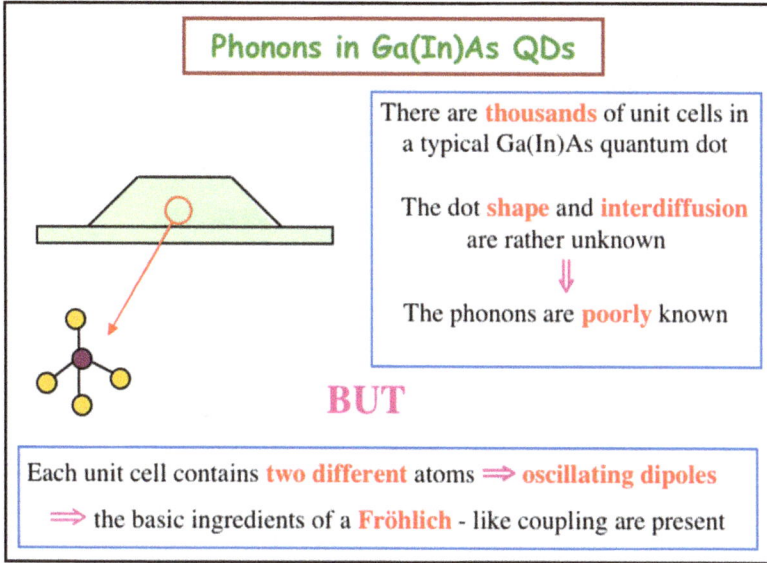

Figure 1.19. Phonons in a (Ga,In)As QD: plausibility of Fröhlich coupling (see text).

modelling based on this simplified description (with a plausible coupling strength) and experiments is an *a posteriori* justification of the approximations made. Note finally that a similar situation occurs in quantum well-based heterostructures, where neglecting the modification of the phonon modes by the material structuration leads to a very fruitful first approach to the description of many phenomena involving the electron–phonon interaction.

We shall consider in this book two cases where the electron–LO interaction plays a crucial role: in the capture (chapter 2) and intra-dot relaxation (chapter 3) processes. For capture, the coupling can be treated at the lowest order approximation, like for free carriers in bulk and quantum wells. In contrast, for intra-dot relaxation the perturbative treatment fails. In order to illustrate this latter point, we will in the next paragraph come back to the initial problem of magneto-transmission experiments and show that all the features presented in figures 1.13–1.16 can be explained by a proper (i.e. beyond perturbation scheme) description of the particular situation in which one bound electron interacts with quasi-monochromatic LO phonons. This proper description introduces the cornerstone for understanding LO phonon-related relaxation (to be discussed in chapter 3) in a QD: the QD polaron framework.

1.5.8 Non-perturbative Fröhlich coupling in quantum dots

With the coupling Hamiltonian being specified (equation (1.29)), we can now switch to the quantum mechanical calculations. The full Hamiltonian is

$$H = H_{QD}(B) + H_{ph} + H_{e-ph} \tag{1.31a}$$

$$H_{ph} = \sum_{\vec{q}} a_{\vec{q}}^{+} a_{\vec{q}} \, \hbar\omega_{LO}(\vec{q}), \tag{1.31b}$$

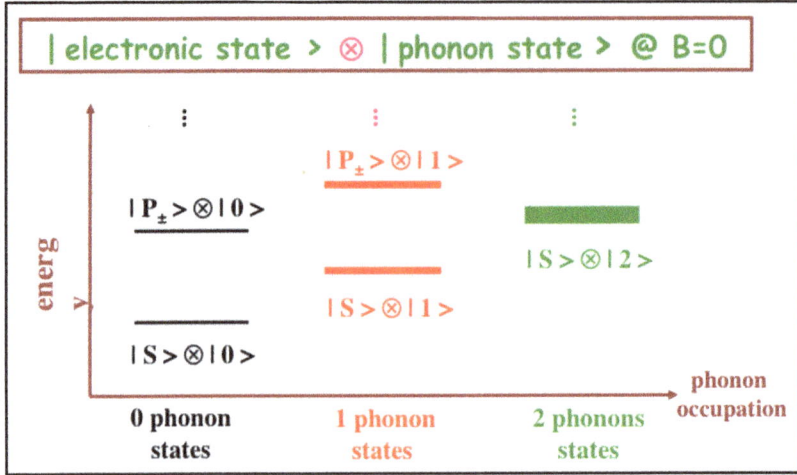

Figure 1.20. The decoupled states of an axial dot (at $B = 0$) retained in the theoretical analysis of the polaron states. The thickness of each level is associated with its phonon width (vanishing for zero-phonon and largest for two-phonon states).

from equations (1.20), (1.28) without the phonon vacuum term and (1.29). Figure 1.20 shows the basis of the decoupled electron–phonon states at $B = 0$ we shall be using. Since the experiments were performed at low temperatures (4–20 K), and considering the high phonon energy ($\hbar\omega_{LO} \approx 36$ meV $\gg k_B T$), it is sensible to retain decoupled states that contain few phonons (0, 1 or 2).

The QD is assumed to display a cylindrical symmetry. $|0\rangle$ corresponds to the phonon vacuum where the atoms undertake zero point vibrations around their equilibrium positions. This phonon state is unique. The $|1\rangle$ and $|2\rangle$ levels represent the continua corresponding to the excitation of one and two arbitrary quanta of vibration, respectively. However, since the LO dispersion is weak (see figure 1.17), these phonon continua are rather flat.

The magnetic field dependences of the energies of the decoupled states result from their electronic components. They are sketched in figure 1.21. The variations of the three zero phonon states are the same as in figure 1.12. We see that all states with the same orbital motion have the same field variation and are separated by $\approx\hbar\omega_{LO}$. We note that crossings (signalled by full circles) show up between the zero-phonon level and the one- or two-LO phonon continuum, and between the one- and two-phonon continua. This is a result of the very different field dependences of the S and $P_\pm$ states as well as the comparable magnitudes of the S–P splitting and LO phonon energy.

The Fröhlich Hamiltonian H_{e-ph} introduces couplings amongst different basis states: in fact, the diagonal couplings vanish and, more generally, only states differing by one LO phonon couple (since H_{e-ph} is linear in the phonon creation and annihilation operators). Hence, we shall move from the physics of the decoupled states where the electrons and phonons evolve independently to an intricate world where some eigenstates are linear superpositions (or entanglements) of several electron states and several phonon occupations: these admixtures are called polarons.

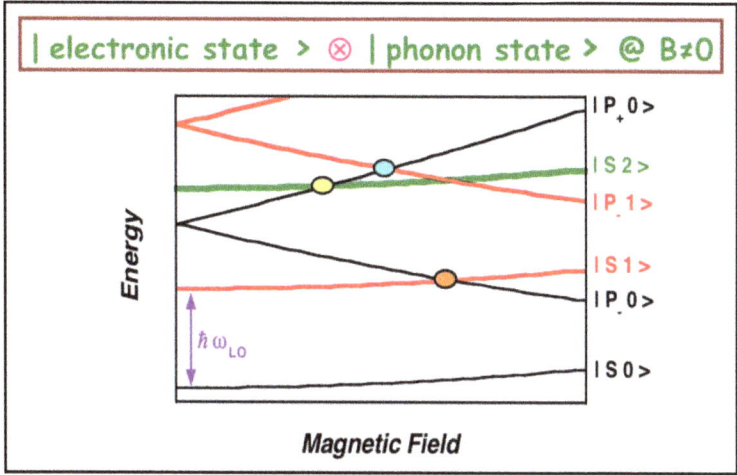

Figure 1.21. Variation with B of the energies of the decoupled states of of an axial dot. The variation of the zero phonon states (black thinner lines) are the same as in figure 1.12.

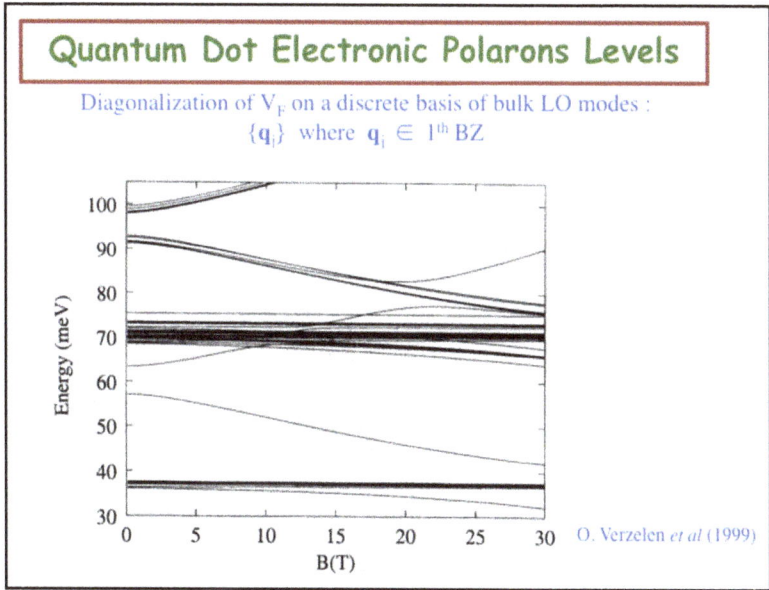

Figure 1.22. Calculated magnetic field dependences of the polaron energies in an anisotropic (Ga,In)As/GaAs self-organized QD. The energy zero has been taken on the $|S,0\rangle$ level. Adapted from Verzelen *et al* (2000).

Figure 1.22 shows the B dependences of the magneto-polaron levels in a (Ga,In) As/GaAs QD. In this numerical calculation, the 3D phonon modes $|1_{\vec{q}}\rangle$ have been discretized within the first Brillouin zone and the LO phonon populations are $n = 0$, 1 or 2. The electron states $|\psi_{1m}\rangle$ retained in the analysis are $|S\rangle$ and $|P_{\pm}\rangle$, and a QD anisotropy has been retained in the modelling. The eigenstates have the generic form

$$|\Psi\rangle = \sum_{m=0,\pm 1} |\psi_{1m}\rangle \otimes \left(C_{m,0} |0\rangle + \sum_{\vec{q}} C_{m,\vec{q}} |1_{\vec{q}}\rangle + \sum_{\vec{q},\vec{Q}} C_{m,\vec{q},\vec{Q}} |1_{\vec{q}}, 1_{\vec{Q}}\rangle \right), \quad (1.32)$$

where the C coefficients result from the numerical diagonalization of H_{e-ph} in the (large) basis of decoupled 0, 1 and 2 phonon states. We easily recognize in figure 1.22 several of the features already present in the decoupled scheme in figure 1.21. However, it is worth stressing that the particular B fields where two ensembles of decoupled states were crossing (circles in figure 1.21) now present anticrossing states. Otherwise, the majority of states behave much like decoupled states. Actually, by analyzing the result of the numerical diagonalization, one essentially finds two classes of eigenstates: (a) the vast majority of states are almost completely non-admixed—projecting almost exclusively on a single electron state and on a single phonon occupancy (i.e. are either zero-phonon, or one-phonon or two-phonon states); (b) in contrast, the few states that anticross present important admixtures of different orbital and phonon states—these are the genuine QD polaron states. The ultimate reason for the existence of two classes of states will become clearer below, when we more thoroughly describe the polaron admixtures within a simplified framework. For the moment, let us show how the numerical study is used to interpret the magneto-optical measurements.

1.5.9 Optical absorption by quantum dot polarons

In order to interpret the magneto-transmission spectra, we should compute the optical absorption probabilities at the dipolar (long wavelength: $\lambda \gg R$) approximation. When considering the matrix element of the electronic dipole between two wavefunctions resulting from the previous diagonalization procedure, it is important to realize that the dipolar coupling V_{dip} acts only on the electronic part and thus that the phonon occupation should be conserved,

$$\langle \Psi_b | V_{dip} | \Psi_a \rangle = \sum_{m_a,m_b} \langle \psi_{1m_b} | V_{dip} | \psi_{1m_a} \rangle P_{m_b,m_a}$$

$$P_{m_b,m_a} = C^*_{m_b,0} C_{m_a,0} + \sum_{\vec{q}} C^*_{m_b,\vec{q}} C_{m_a,\vec{q}} + \sum_{\vec{q},\vec{Q}} C^*_{m_b,\vec{q},\vec{Q}} C_{m_a,\vec{q},\vec{Q}}. \quad (1.33)$$

This significantly restricts the number of allowed transitions. Nevertheless, recall that the light also directly interacts with the *transverse* optical modes, which are responsible for the strong absorption band around ≈ 35 meV in figures (1.13) and (1.15), but do not Fröhlich couple to the electrons.

Since the ground state $|S,0\rangle$ is well separated from any other state (see, for example, figure 1.21), it is the only one to be thermally populated at $T = 4$–10 K; it has very little admixture with the other high-energy decoupled states ($C_{0,0} \approx 1$). To be efficiently light-coupled to this initial state, the high-energy ones should display a significant projection on $|P_\pm,0\rangle$ (figure 1.23).

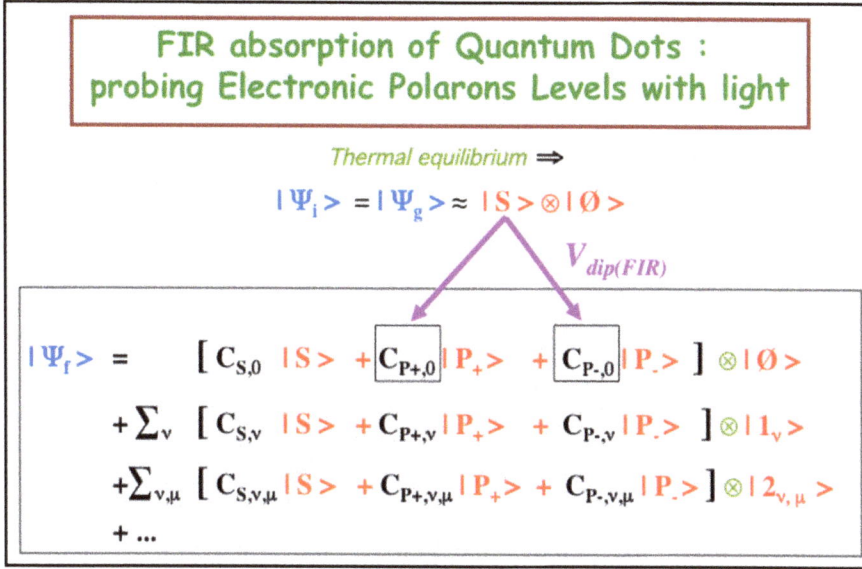

FIR absorption of Quantum Dots : probing Electronic Polarons Levels with light

Thermal equilibrium $\Rightarrow$

$$|\Psi_i> = |\Psi_g> \approx |S> \otimes |\emptyset>$$

$$V_{dip(FIR)}$$

$$|\Psi_f> = [\, C_{S,0}\, |S> + C_{P+,0}\, |P_+> + C_{P-,0}\, |P_-> \,] \otimes |\emptyset>$$

$$+ \sum_{\nu} [\, C_{S,\nu}\, |S> + C_{P+,\nu}\, |P_+> + C_{P-,\nu}\, |P_-> \,] \otimes |1_\nu>$$

$$+ \sum_{\nu,\mu} [\, C_{S,\nu,\mu}\, |S> + C_{P+,\nu,\mu}\, |P_+> + C_{P-,\nu,\mu}\, |P_-> \,] \otimes |2_{\nu,\mu}>$$

$$+ \ ...$$

Figure 1.23. Initial and final magneto-polaron states projected on the decoupled basis. The arrows indicate the components coupled by the dipolar coupling, considering both phonon number (conserved) and orbital angular momentum (varied by ±1) conservation rules.

Despite the huge number of states in the basis, selection rules on the phonon number and orbital angular momentum severely restrict the number of efficient transitions, and magneto-absorption experiments finally give access to only a small number of magneto-polaron states. Strikingly, these are the states with the more marked polaron admixing (class (b) in section 1.5.8) that contribute most to the optical signal. Indeed, figure 1.24 shows a comparison between the experimental results (symbols) and the calculated transitions towards the whole ensemble of diagonalized final states, but where only those with 'intensity' $|\langle \Psi_f | V_{dip} | \Psi_i \rangle|^2$ larger than 10% of the strongest one are represented.

It turns out that in the vicinity of an anticrossing the absorption is governed by the sole states that anticross. This is particularly the case for the $|S,1\rangle$–$|P_-,0\rangle$ anticrossing near 30 T, and for the one near 24 T between the 'ascending' $|P_+,0\rangle$ and 'descending' $|P_-,1\rangle$ branches. The small anticrossing of the 'ascending' branch near 12 T is in reality a 'frustrated' anticrossing since $|P_+,0\rangle$ and $|S,2\rangle$ have the same energy but are not directly coupled by the Fröhlich Hamiltonian, which is linear in phonon creation and annihilation and therefore only couples phonon occupations that differ by ±1.

In conclusion, magneto-transmission measurements in doped QDs are a very selective probe of the 'true' QD polaron states: namely, the ones that anticross when put into resonance by the magnetic field and for which the wavefunctions are entanglements of electron states with different orbital motions and phonon states with different occupations.

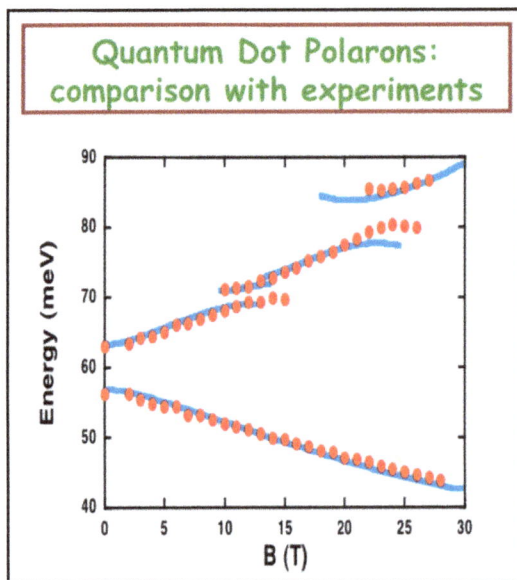

Figure 1.24. Calculated (blue solid lines) and measured (symbols) magneto-optical transitions in a self-assembled QD. Adapted from Hameau *et al* (1999).

As shown in figure 1.25, the strong coupling regime between electrons bound to QDs and LO phonons has been observed in several samples corresponding to dots of different mean sizes, but all with an average QD occupation of one electron. From left to right panels, the bare S–P energy increases (QDs with smaller in-plane dimensions). However, the fan charts are much the same except that the one-phonon and two-phonon anticrossings on the 'ascending' branch exchange their positions: the two-phonon anticrossing appears at the lower (higher) field on the right (left) panel.

Let us finally briefly mention evidence for polaron couplings for holes in p-doped (Preisler *et al* (2005)) and electron–hole pairs (or excitons; Verzelen *et al* (2002) and Preisler *et al* (2006)) in intrinsic QDs, which confirm the rather universal strong coupling between particles or particle complexes bound to QDs and LO phonons.

1.5.10 Polaron features in a quantum dot ensemble

As mentioned above, one of the major advantages of magneto-absorption is that it provides a tuneable control parameter allowing decoupled levels to be lined up, or conversely their detuning. Optical absorption experiments undertaken at $B = 0$ on an inhomogeneous ensemble of QDs do not allow such an unambiguous identification of polaron effects in QDs. Nevertheless, when the lateral sizes of the QDs change one may induce a sizeable polaron effect. This is shown in figure 1.26 where the calculated polaron states are plotted versus the radius of the axial QD (at $B = 0$). We note a pronounced anticrossing near 130 Å, which corresponds to the two polaron branches built out of the interacting $|S,1\rangle$ and $|P_\pm,0\rangle$. The energy separation

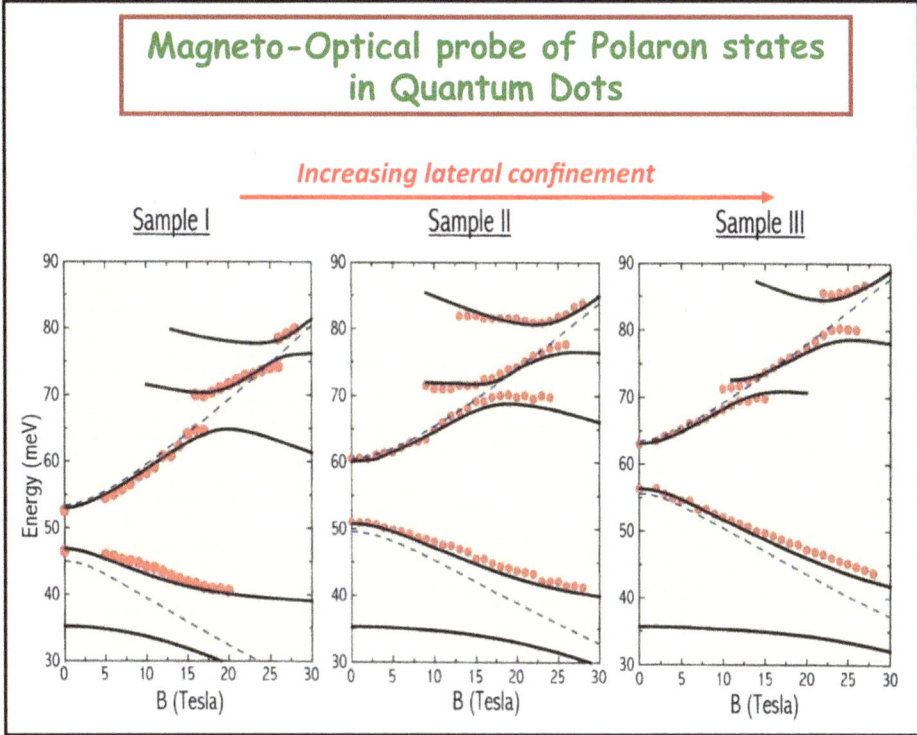

Figure 1.25. Calculated (solid black lines) and measured (symbols) magneto-optical transitions in (Ga,In)As self-assembled QDs that display an increasing spatial confinement. The QDs are modulation-doped with one electron on average per QD. The dashed lines correspond to the anisotropic macro-atom model (purely electronic excitations). Adapted from Hameau *et al* (2002).

at resonance ($\approx$12 meV) is bigger than the broadening of the S–P transition ($\approx$5 meV in figures 1.13 and 1.15), so that the electronic polaron effect is robust against the inhomogeneous size broadening of the dot ensemble (note that in this figure the 1S-related levels in the inset are denoted by a tilde and are slightly energy shifted as compared to the bare ones $|S0\rangle$ and $|S1\rangle$, separated by $\hbar\omega_{LO}$).

We should finally stress the very particular situation of the (Ga,In)As/GaAs QDs, where self-assembly means that the S–P transition energy is of the same order of magnitude as the LO phonon energy, thereby leading to pronounced polaron effects.

1.5.11 Analytical model for the quantum dot polarons

Numerical computations provide us with a very small number of entangled electron–phonon states that are optically coupled to the ground state, in agreement with the measured spectra (see sections 1.5.8 and 1.5.9). A simple model of a discrete level coupled to a continuum with finite width allows a better understanding of these results. We assume that an external parameter (like the magnetic field) can push the uncoupled discrete state inside and outside the continuum (top scheme in figure 1.27,

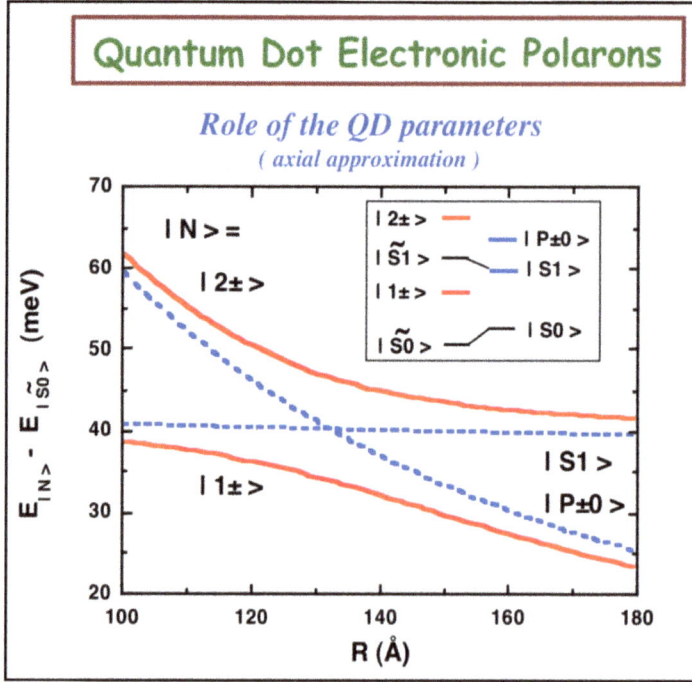

Figure 1.26. Calculated decoupled states (dashed lines) and polaron energies (full lines), measured from the energy of the dressed ground state $|\tilde{S},0\rangle$ (not shown), versus the basal radius of the truncated cone ($h/R = 0.15$ and $d = 3.33$ Å in figure 1.2). Inset: scheme of polaron (left) and decoupled (right) states. From Verzelen *et al* (2000).

corresponding with one of the *B*-induced resonances in figure 1.21). To describe the interacting system, the key parameter is the dimensionless ratio,

$$r = \frac{\langle\langle |V|\rangle\rangle}{\Delta_c}, \tag{1.34}$$

between the averaged $\langle\langle |V|\rangle\rangle$ coupling strength between the discrete state and the continuum (viz. the average over the continuum states of the absolute value of the coupling matrix element) to the width Δ_c of the continuum (see, for example, Cohen-Tannoudji (1996)). When $r \ll 1$ the coupled system (discrete state plus continuum) is in a weak coupling regime, and the discrete state irreversibly dilutes in the continuum when pushed inside it by the external parameter (bottom-left scheme in figure 1.27). In contrast, if $r \gg 1$ the discrete state does not enter the continuum: instead, two branches are formed that end on both sides of the continuum (bottom-right scheme in figure 1.27). If $r \gg 1$ the coupling is sufficiently strong to 'reorganize' the continuum: this reorganization amounts to creating an 'average or effective continuum level' to which the discrete state couples preferentially. The general situation can be tackled in the framework of the famous Fano model (as applied to a finite-width continuum). However, the solution of the interacting problem is

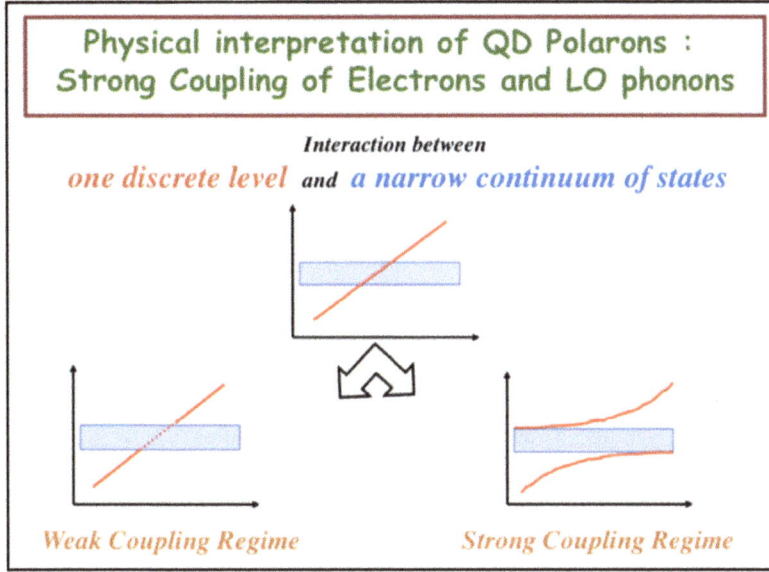

Figure 1.27. The two possible fates of a discrete level made to cross a narrow continuum (top scheme). If $r \ll 1$, weak coupling is reached and the discrete state irreversibly dissolves in the continuum (bottom-left). If $r \gg 1$ the system is in the strong coupling regime where there is an anticrossing behaviour, with two branches that end at both boundaries of the continuum with a finite width.

particularly simple in the $r \to \infty$ limit of a very narrow continuum, where closed expressions can be readily obtained for the energies, the admixed eigenstates and also for the 'effective' state with which the discrete one is coupled. We present in the following a description of the QD polaron eigenstates and of the 'effective one-phonon state' in the $r = \infty$ limit. The pertinence of this limit for QD polarons is to a large extent due to the fact that continuum broadening is in this case due to phonon dispersion, and that the already small LO phonon dispersion width $\Delta_{LO} \approx 8$ meV in figure 1.17 is actually much smaller in practice, since only phonons with small wavevectors (typically $q < \pi/$(QD size)) do actually couple to the bound electron and contribute to $\langle\langle|V|\rangle\rangle$ in equation (1.34).

In the QD polaron problem, the one- (or more) phonon continuum becomes flat when LO phonon dispersion is neglected. Let us develop an analytical model of the one-phonon QD polaron in this regime. However, for the sake of generality, let us retain for a while the true phonon dispersion $\hbar\omega_{LO}(\vec{q})$. We search then for the eigenstates of the total Hamiltonian in equation (1.31). To this end, we consider the ensemble $\{\vec{q}_n\}$ of N phonon wavevectors resulting from their quantization in a very large 3D box. To introduce the model we focus our attention only on the 'descending' branch in figure 1.21, and search for a solution in the form

$$|\psi\rangle = a\,|1P_-, 0\rangle + \sum_{\vec{q}_n} b(\vec{q}_n)\big|1S, 1_{\vec{q}_n}\big\rangle. \tag{1.35}$$

We get an $(N + 1) \times (N + 1)$ eigenvalue problem to solve:

$$\begin{vmatrix} E_{P_-} - \varepsilon & \lambda_1 & \lambda_2 & \dots & \lambda_N \\ \lambda_1^* & E_S + \hbar\omega(\vec{q}_1) - \varepsilon & 0 & \dots & 0 \\ \lambda_2^* & 0 & E_S + \hbar\omega(\vec{q}_2) - \varepsilon & \dots & 0 \\ \dots & \dots & \dots & \dots & \dots \\ \lambda_N^* & 0 & 0 & \dots & E_S + \hbar\omega(\vec{q}_N) - \varepsilon \end{vmatrix} \begin{vmatrix} a \\ b(\vec{q}_1) \\ b(\vec{q}_2) \\ \dots \\ b(\vec{q}_N) \end{vmatrix} = \begin{vmatrix} 0 \\ 0 \\ 0 \\ \dots \\ 0 \end{vmatrix},$$

$$(1.36a)$$

where

$$\lambda_p \equiv \langle 1P_-; 0 \mid H_{\text{e–ph}} \mid 1S; 1_{\vec{q}_p} \rangle = -\mathrm{i}\frac{C_{\text{F}}}{q_p\sqrt{\Omega_{\text{cr}}}}\langle 1P_- \mid \mathrm{e}^{\mathrm{i}\vec{q}_p \cdot \vec{r}} \mid 1S \rangle. \qquad (1.36b)$$

Equation (1.36a) can equivalently be rewritten as

$$\begin{cases} (E_P - \varepsilon)a + \sum_{n=1}^{N}\lambda_n \, b(\vec{q}_n) = 0 \\ \left(E_S + \hbar\omega_{\text{LO}}(\vec{q}_n) - \varepsilon\right)b(\vec{q}_n) + \lambda_n^* a = 0 \end{cases} \qquad (1.37a)$$

$$\Rightarrow E_P - \varepsilon = \sum_{n=1}^{N}\frac{|\lambda_n|^2}{E_S + \hbar\omega_{\text{LO}}(\vec{q}_n) - \varepsilon}, \qquad (1.37b)$$

Note that the denominators should never vanish in equation (1.37b). In the limit of a flat LO phonon continuum (no LO phonon dispersion: $\hbar\omega_{\text{LO}}(\vec{q}) \approx \hbar\omega_{\text{LO}}(0) \approx \hbar\omega_{\text{LO}}$), we obtain two polaron solutions that fulfil

$$(E_{P_-} - \varepsilon)(E_S + \hbar\omega_{\text{LO}} - \varepsilon) = \sum_{n=1}^{N}|\lambda_n|^2 \equiv |V_{\text{eff}}|^2$$

$$\Rightarrow \varepsilon_\pm = \frac{1}{2}[E_{P_-} + E_S + \hbar\omega_{\text{LO}}] \pm \frac{1}{2}\sqrt{(E_S + \hbar\omega_{\text{LO}} - E_{P_-})^2 + 4|V_{\text{eff}}|^2}. \qquad (1.38)$$

These two solutions exhibit the familiar energy repulsion of interacting quantum states. Indeed, we find readily that the + (−) root is at larger (smaller) energy than the higher (lower) of E_{P_-} and $E_S + \hbar\omega_{\text{LO}}$. We note that these two solutions are entanglements of different electronic levels with different phonon occupations: they can never be factorized in terms of an electronic state times a definite phonon occupation (the eigenfunctions related to the energies $\varepsilon_\pm$ are presented later in section 1.5.13).

As we stated before, equation (1.38) holds only if the energy denominators of equation (1.37b) do not vanish. But then equation (1.38) gives us only two solutions. The N–1 missing solutions are in fact scrambled unadmixed LO phonon states; that is to say, these states are factorizable in terms of an electronic state $|1S\rangle$ times a linear superposition of one-LO phonon states. Indeed, if $\varepsilon = E_S + \hbar\omega_{\text{LO}}$, we have

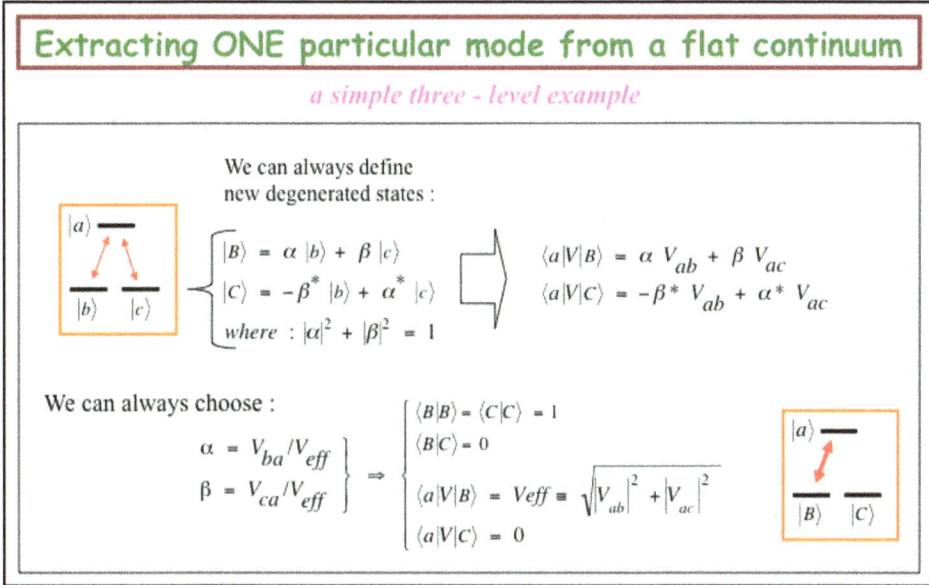

Figure 1.28. Extracting one particular mode from a degenerate continuum: a simple three-level example.

$a = 0$ from the second equation (1.37a). For $a = 0$ the first equation (1.37a) still holds, provided that $\sum_{n=1}^{N} \lambda_n \, b_\nu(\vec{q}_n) = 0$ is fulfilled, where ν labels its different solutions. Since there is always one linear relation that has to be fulfilled per solution, only $N{-}1$ are free. Since $a = 0$, they all correspond to one-LO phonon replicas. Hence, we recover the right number of possible solutions: $(N{-}1) + 2$.

The two entangled states corresponding to the energies $\varepsilon_\pm$ that have been obtained for a 'monochromatic' continuum ($\Delta_c = 0$) reflect the reorganization of this flat continuum due to its coupling to the discrete state. Indeed, due to the degeneracy of the N unperturbed levels $|1S\rangle \otimes |1_{\vec{q}_n}\rangle$, any linear superposition of these states has the same energy $E_S + \hbar\omega_{LO}$. It can actually be shown that it is possible to generate a unique linear superposition of $|1S\rangle \otimes |1_{\vec{q}_n}\rangle$ states that will be coupled to the discrete states, and such that any other linear superposition orthogonal to it actually will not be coupled to the discrete state. We illustrate this feature first in a simplistic three-level system: see figure 1.28. In this figure, a well-chosen linear combination of the degenerate states $|b\rangle$ and $|c\rangle$ generates states $|B\rangle$ and $|C\rangle$ which are orthonormal and such that only $|B\rangle$ is coupled to the discrete state $|a\rangle$, with an effective coupling that conserves the total interaction strength: $|V_{aB}|^2 = |V_{\text{eff}}|^2 = |V_{ab}|^2 + |V_{ac}|^2$.

This procedure can be easily generalized to the flat phonon continuum, yielding the single one-phonon mode

$$|1_{SP}\rangle = \frac{1}{V_{\text{eff}}} \sum_{\vec{q}_n} \lambda_n^* |1_{\vec{q}_n}\rangle; \quad V_{\text{eff}} = \sqrt{\sum_{\vec{q}_n} |\lambda_n|^2}, \qquad (1.39)$$

with λ_n given in equation (1.36b). It is then easy to show that any other one-phonon linear combination $|1_\nu\rangle = \sum_{\vec{q}_n} b_{\nu,n} |1_{\vec{q}_n}\rangle$ orthogonal to $|1_{SP}\rangle$, i.e. for which $\sum_{\vec{q}_n} b_{\nu,n}^* \lambda_n^* = 0$, cannot contribute to the formation of QD polaron states,

$$\left\langle 1S, 1_\nu \right| H_{\text{e-ph}} |1P_-, 0\rangle = \sum_{\vec{q}_n} b_{\nu,n}^* \left\langle 1S, 1_{\vec{q}_n} \right| H_{\text{e-ph}} |1P_-, 0\rangle \equiv \sum_{\vec{q}_n} b_{\nu,n}^* \lambda_n^* = 0. \quad (1.40)$$

Thus, for a flat continuum ($r \to \infty$), the problem of N levels coupled to one discrete state amounts to solving a two-level problem with an effective coupling V_{eff}, while the N–1 other levels show no admixture to the discrete state.

It is now easy to convince ourselves that this same formalism can be further generalized to incorporate the two $|P_\pm\rangle$ and the two-phonon flat continuum. However, note that the two extracted one-phonon states are independent,

$$\langle 1_{SP_+} | 1_{SP_-} \rangle = \frac{1}{V_{\text{eff}}^2} \sum_{\vec{q}_n} \lambda_{-,n} \lambda_{+,n}^* = \frac{C_F^2}{\Omega_{\text{cr}} V_{\text{eff}}^2} \sum_{\vec{q}_n} \langle P_-| e^{i\vec{q}_n \cdot \vec{r}} |S\rangle\langle S| e^{-i\vec{q}_n \cdot \vec{r}} |P_+\rangle$$

$$= \frac{C_F^2}{(2\pi)^3 V_{\text{eff}}^2} \int d\vec{r} \left| \psi_{1S}(\vec{r}) \right|^2 \psi_{1,-1}^*(\vec{r})\psi_{1,+1}(\vec{r}) = 0, \quad (1.41)$$

where we used $\sum_{\vec{q}} \exp\{i\vec{q} \cdot (\vec{r}_1 - \vec{r}_2)\} = \Omega_{\text{cr}}\delta(\vec{r}_1 - \vec{r}_2)/(2\pi)^3$ and the angular part of the last integral vanishes on account of the $\exp\{im\varphi\}$ dependences of the ψ_{1m} states. Note also that the two-phonon excitations to be retained in the study of the inter-shell S–P polarons are necessarily of the form $|2_{SP_+}\rangle, |2_{SP_-}\rangle$ and $|1_{SP_+}; 1_{SP_-}\rangle$, so that the working basis is extremely reduced in this model as compared to the one used to obtain figure 1.22.

We shall not develop this model any further, but only mention that the computations of transition energies (and also line intensities) in self-assembled QDs using the simplest scheme of a degenerate phonon continuum are in excellent agreement with the numerical simulations and therefore with measurements.

Let us point out the early numerical work of Inoshita and Sakaki (1997), who studied via a Green function method the density of states for the interacting electron–LO phonon system in a dot. They proved the existence of polaron splitting of the energy levels when the ground-to-first excited inter-level separation of their model QD was made to be resonant with the dispersionless LO phonon energy. Their numerical method, however, did not allow them to give explicit analytical expressions of the characteristics of the polaron eigenstates, as was presented in this chapter.

Before concluding, it is worth pointing out the analogy between the electrons bound to QDs interacting with almost dispersionless LO phonons, and the radiative recombination of excited electrons of an atom. In the latter situation, if the atom is in vacuum (or a macroscopic cavity) the initial state is the discrete state $|e\rangle \otimes |0_{\text{phot}}\rangle$, which is embedded in the continuum $|g\rangle \otimes |1_{\vec{q}}\rangle$ formed between the atom in the ground state $|g\rangle$ and one emitted photon with wavevector $\vec{q}$. The discrete state and the continuum are coupled by the electric dipole $\langle e|\vec{r}|g\rangle$. It is well known that the discrete state dissolves in the continuum and that the de-excitation of an atom is irreversible. Now, if the radiation is imprisoned in a cavity, the modes of the electromagnetic field are no longer continuously distributed but become discrete (cavity quantum electrodynamics): $\vec{q} \to \vec{q}_{l,m,n}$. As a result, it has been shown that the atom undertakes cycles of emission/absorption and oscillates in time between $|e\rangle$ and $|g\rangle$. The correct elementary excitations are no longer the factorized states but the cavity polaritons whose wavefunctions are entanglements of different factorized states: $|\Psi\rangle = \alpha_0 |e\rangle \otimes |0_{\text{phot}}\rangle + \sum_{l,m,n} \alpha_{l,m,n} |g\rangle \otimes |1_{q_{l,m,n}}\rangle$.

1.5.12 Localized phonon mode

The analytical model developed in the previous section to describe the Fröhlich coupling between a discrete state and a flat continuum considerably simplifies the numerical computation of several physical observables of interest. Besides the magneto-absorption spectrum, one can evaluate various properties of the particular phonon mode $|1_{\text{SP}}\rangle$. Let us briefly consider (without providing the analytical details) the average oscillation amplitude of the atoms that are involved in the formation of the polaron states resulting from the coupling of $|P_-, 0\rangle$ and $|S, 1_{\vec{q}}\rangle$.

Figure 1.29 shows the variation (with respect to the bulk value) of this amplitude versus the distance d of the lattice cell to the QD centre, keeping the altitude z constant (at the basis of the QD, right on top of the WL). We note that only the atoms inside the QD participate in this particular LO vibration mode $|1_{\text{SP}}\rangle$. Also, the amplitude vanishes on the symmetry Oz axis, i.e. when the P-state wavefunction

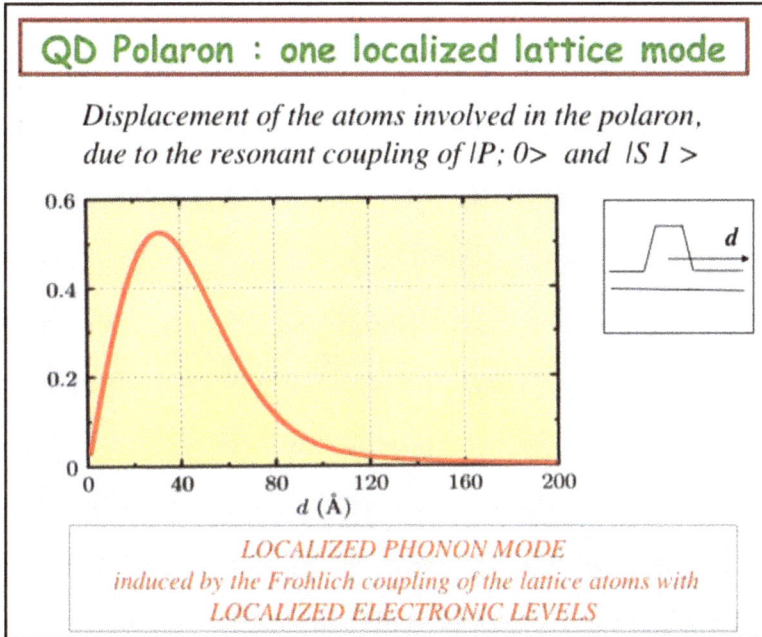

Figure 1.29. Amplitude (arbitrary units) of the localized phonon mode $|1_{\text{SP}}\rangle$ induced by the Fröhlich coupling of the lattice atoms with electron states bound to the QD. $R = 10$ nm.

vanishes. For a propagating phonon mode, the same calculation provides a constant amplitude. Thus, a localized vibration mode (phonon wavepacket) results from the interaction of the propagating phonons with the confined electron states.

1.5.13 Dynamical aspects of the polaron states

For dispersionless LO phonon modes, the polaron states formed from the coupling of $|P, 0\rangle$ and $|S, 1_{SP}\rangle$ are solutions of

$$|\psi(t)\rangle = a(t)|P; 0_{SP}\rangle + b(t)|S; 1_{SP}\rangle$$

$$\Rightarrow i\hbar \frac{\partial}{\partial t}\begin{pmatrix} a \\ b \end{pmatrix} = \begin{pmatrix} E_P & V_{eff} \\ V_{eff}^* & E_S + \hbar\omega_{LO} \end{pmatrix}\begin{pmatrix} a \\ b \end{pmatrix}. \tag{1.42}$$

The two stationary solutions are the polaron states

$$\left|\psi_+(t)\right\rangle = |2\rangle \exp(-i\varepsilon_+ t/\hbar)$$

$$\left|\psi_-(t)\right\rangle = |1\rangle \exp(-i\varepsilon_- t/\hbar)$$

$$\begin{cases} |2\rangle = \cos(\theta)|P; 0_{SP}\rangle + \sin(\theta)|S; 1_{SP}\rangle \\ |1\rangle = -\sin(\theta)|P; 0_{SP}\rangle + \cos(\theta)|S; 1_{SP}\rangle \end{cases} \tag{1.43}$$

$$\sin^2(\theta) = \frac{1}{2}\left\{ 1 - \frac{\Delta E}{\sqrt{\Delta E^2 + 4V_{eff}^2}} \right\},$$

with energy detuning given by $\Delta E = E_P - (E_S + \hbar\omega_{LO})$ and where we have used the notation $|2\rangle$ ($|1\rangle$) for the upper (lower) time-independent polaron eigenstate (see, for example, figure 1.26). The general solution of the time-dependent problem is

$$\left|\psi(t)\right\rangle = \alpha\left|\psi_+(t)\right\rangle + \beta\left|\psi_-(t)\right\rangle$$

$$\Rightarrow \begin{cases} a(t) = \alpha\cos(\theta)\exp(-i\varepsilon_+ t/\hbar) - \beta\sin(\theta)\exp(-i\varepsilon_- t/\hbar) \\ b(t) = \alpha\sin(\theta)\exp(-i\varepsilon_+ t/\hbar) + \beta\cos(\theta)\exp(-i\varepsilon_- t/\hbar), \end{cases} \tag{1.44}$$

where α and β are two constants such that $|\alpha|^2 + |\beta|^2 = 1$. If $|\psi(t = 0)\rangle = |P; 0_{SP}\rangle$, then $\alpha = \cos(\theta)$, $\beta = -\sin(\theta)$, and the probability to find the system in this zero-phonon state (survival probability) is

$$P_0(t) \equiv \left|\langle\psi(0)|\psi(t)\rangle\right|^2 = \left|a(t)\right|^2 = 1 - \sin^2(2\theta)\sin^2\left[\sqrt{(\Delta E/2)^2 + V_{eff}^2}\, t/\hbar\right]. \tag{1.45}$$

The time-evolution appears simply because $|\psi(t = 0)\rangle$ is not an eigenstate of the full Hamiltonian (including electron–phonon coupling). In particular at resonance there is $\theta = \pm\pi/4$ and $P_0(t) = \cos^2[V_{eff}\, t/\hbar]$: the system oscillates between the zero-phonon and one-phonon components, with frequency $V_{eff}/\hbar$. This is the dynamical counterpart of the existence of admixed stationary polaron states: if one populates at $t = 0$ the independent (zero-phonon) state of the dot, then Fröhlich coupling makes the system oscillate periodically between the two basis states: from the phonon point of view, this corresponds to periodic emission and re-absorption of one $|1_{SP}\rangle$ LO phonon, while from the electron side this corresponds to a periodic orbital changing. As we will see in the last chapter, the re-absorption process is no longer complete when the LO phonon mode acquires a finite lifetime, leading in the end to an irreversible orbital changing for the electron from the P towards the S shell.

References

Bastard G 1988 *Wave Machanics Applied to Semiconductor Heterostructures* (Paris: Les Editions de Physique)

Bastard G, Brum J A and Ferreira R 1991 Solid state physics: Semiconductor Heterostructures and Nanostructures *Electronic States in Semiconductor Heterostructures* ed H Ehrenreich and D Turnbull (Berlin: Springer)

Bimberg D, Grundmann M and Ledentsov N N 1999 *Quantum Dot Heterostructures* (New York: Wiley)

Cohen-Tannoudji C, Dupont-Roc J and Grynberg G 1996 *Processus d'interaction antre photons et atomes* (Paris: Les Editions du CNRS)

Hameau S, Guldner Y, Verzelen O, Ferreira R, Bastard G, Zeman J, Lemaître A and Gérard J M 1999 Strong electron–phonon coupling regime in quantum dots: evidence for everlasting resonant polaron *Phys. Rev. Lett.* **83** 4152

Hameau S, Isaia J N, Guldner Y, Deleporte E, Verzelen O, Ferreira R, Bastard G, Zeman J and Gérard J-M 2002 Far-infrared magnetospectroscopy of polaron states in self-assembled InAs/GaAs quantum dots *Phys. Rev.* B **65** 085316

Inoshita T and Sakaki H 1997 Density of states and phonon-induced relaxation of electrons in semiconductor quantum dots *Phys. Rev.* B **56** R4355

Landölt-Bornstein Group III Condensed Matter Series1970–2014 *Numerical Data and Functional Relationships in Science and Technology* (Berlin: Springer)

Lelong Ph and Bastard G 1996 Excitons and charged excitons binding energies in quantum dots *Solid State Commun.* **98** 819

Marzin J Y, Gérard J-M, Izrael A, Barrier D and Bastard G 1994 Photoluminescence of single InAs quantum dots obtained by self-organised growth on GaAs *Phys. Rev. Lett.* **73** 716

Pang Q, Zhao L, Cai Y, Nguyen D-P, Regnault N, Wang N, Yang S, Ge W, Ferreira R, Bastard J and Wang J 2005 CdSe Nano-tetrapods: Controllable Synthesis, Structure Analysis, and Electronic and Optical Properties *Chem. Mater.* **17** 5263

Rossi F and Zanardi P (ed) 1995 *Semiconductor Macroatoms* (London: Imperial College Press)

Preisler V, Ferreira R, Hameau S, de Vaulchier L A, Guldner Y, Sadowski M L and Lemaitre A 2005 Hole–LO phonon interaction in InAs/GaAs quantum dots *Phys. Rev.* B **72** 115309

Preisler V, Grange T, Ferreira R, de Vaulchier L A, Guldner Y, Teran F J, Potemski M and Lemaître A 2006 Evidence for excitonic polarons in InAs/GaAs quantum dots *Phys. Rev.* B **73** 075320

Ridley B K 1988 *Quantum Processes in Semiconductors* (Oxford: Clarendon)

Schiff L 1968 *Quantum Mechanics* (New York: McGraw-Hill)

Stier O 2001 Berlin studies in solid state physics *Electronic and Optical Properties of Quantum Dots and Wires*

Verzelen O, Ferreira R and Bastard G 2002 Excitonic polarons in Semiconductor Quantum Dots *Phys. Rev. Lett.* **88** 146803

Yu P Y and Cardona M 1999 *Fundamentals of Semiconductors: Physics and Materials Properties* (Berlin: Springer)

IOP Concise Physics

Capture and Relaxation in Self-Assembled
Semiconductor Quantum Dots
The dot and its environment
Robson Ferreira and Gérald Bastard

Chapter 2

Capture of carriers by the quantum dots

The exchanges of energy and particles between a QD and its environment (WL and barriers) play a crucial role in its physical properties. For instance, emitters made of QDs (lasers, diodes) must be continuously fed with carriers, and this is usually done by optically or electrically injecting carriers in the continuum, from where a subsequent capture into the dots occurs. In the same way, doped dots used in infrared detection devices must equally be refed after each photon absorption process that promotes the electron into the continuum.

The capture of one carrier by a QD, as well as its relaxation towards the fundamental bound level (to be discussed in the next chapter), cannot occur without the presence of one or several reservoirs. These reservoirs play a twofold role: to absorb the energy lost by the carrier during the capture and relaxation processes; and to ensure the irreversibility of these processes. Capture and relaxation processes are thus assisted by the reservoir with which the carrier interacts.

The typical experimental situation we will consider is the one with carriers (electrons, holes) injected, either optically or electrically, in the continuum of states of the WL. A few mechanisms have been proposed to explain the capture towards (and relaxation within) the bound states of the dots. We distinguish between single particle and collective processes, according to whether the reservoir acts on individual carriers or is related to the interactions between them.

Figure 2.1 presents different assisted capture and relaxation mechanisms in self-assembled QDs. They imply one, two or a gas of carriers, and different phonon-related reservoirs. The capture assisted by the optical phonons and the one related to the Coulombic interactions among the electrons of a free WL gas, are separately presented and detailed in this chapter. We will consider the rate for the capture of a first carrier, and also the one for the capture of a second carrier by a charged dot. Some of the processes in figure 2.1, such as the capture related to polaron lifetime,

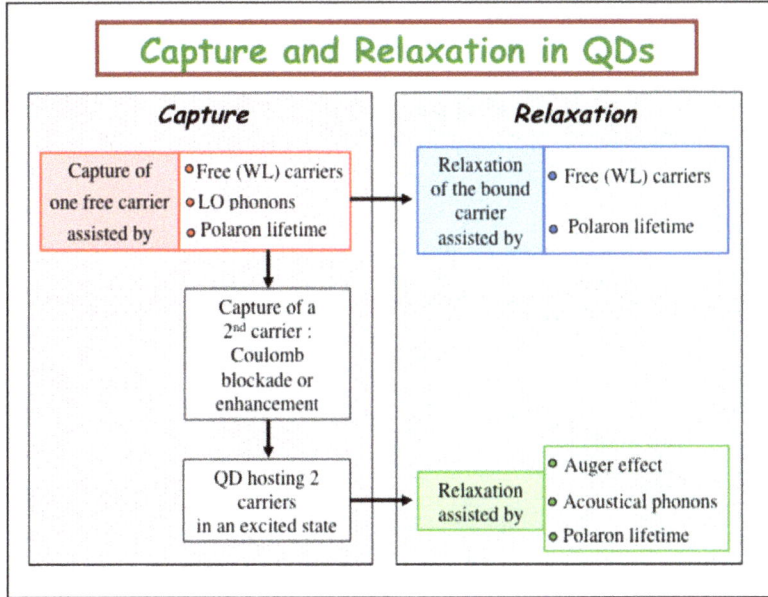

Figure 2.1. Schematic view of various assisted capture and relaxation processes in self-assembled QDs involving one or two carriers.

will not be discussed (see Magnusdottir *et al* (2002)). Finally, the relaxation processes (for a simply or doubly occupied dot) will be discussed in the next chapter.

2.1 Phonon-assisted capture by one empty dot

2.1.1 Energy and size selectivity

Phonons play a crucial role in dot/environment exchanges. Capture can in principle be assisted by the emission of one acoustical or optical phonon. However, energy conservation means that only dots with a 'marginally bound' state (i.e. distant from the continuum by at most the characteristic phonon energy) can benefit from this capture process. Thus, since the energy of the dot bound states (and in particular of the very excited ones) varies rapidly with the dot parameters (see, for example, figure 1.4), this capture mechanism applies only to a fraction of dots of an inhomogeneous ensemble. This constraint is even more restrictive for acoustical phonons, with characteristic energy of a few milli-electron volts, while it is of a few tens of milli-electron volts for optical phonons. Hence, we shall consider only the capture due to coupling between electrons and optical phonons (note however that acoustical phonons become increasingly important at high temperatures).

The Fröhlich interaction was detailed in the previous chapter, in relation to the formation of polaron states as a consequence of the strong coupling between confined electrons and LO phonons. Here, for the capture process, we will consider the same interaction as a perturbation; weak enough so as to leave the electronic states unchanged, but efficient enough to induce scattering of the electron among its different states. As we shall show, the perturbation scheme can be applied owing to

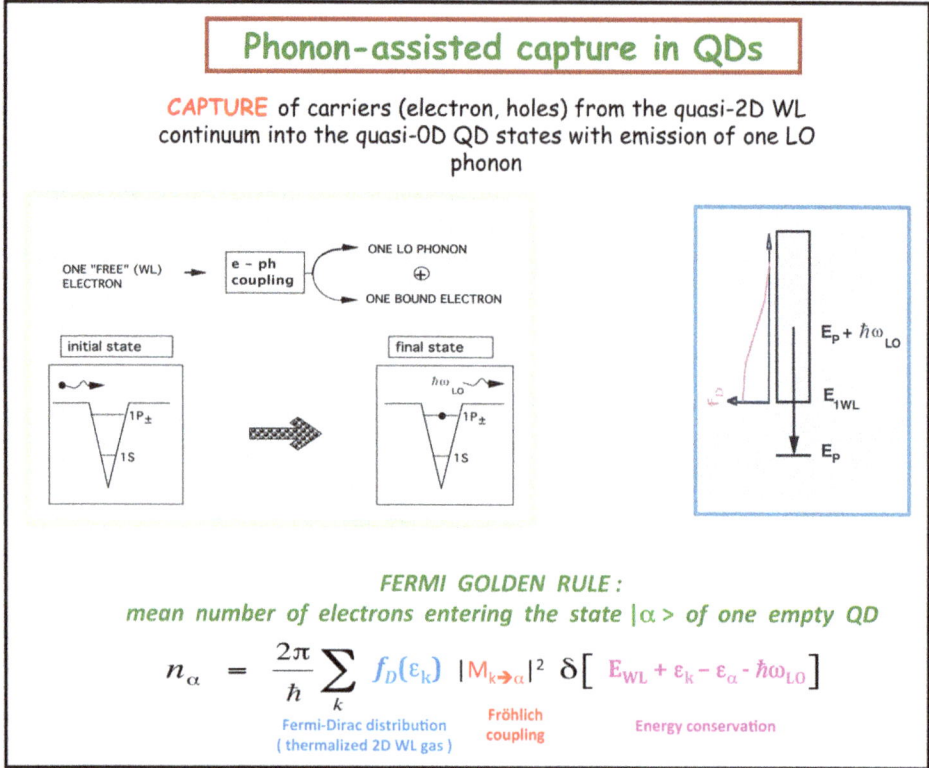

Figure 2.2. Top left: scheme of electron capture with emission of one optical phonon. Top right: illustration of energy conservation for capture from a gas of free carriers into the bound P level.

the existence of a continuous set of initial states in a capture process. Thus, in the following we will consider the Fröhlich coupling H_{e-ph} at the lowest order of perturbation theory (Fermi golden rule) to evaluate the rates for scattering from the populated free states of the WL and towards a fixed bound state of a given dot. A sketch of this process is given in figure 2.2 for the case where the bound state is in the P shell of the QD.

To evaluate the mean number of electrons entering per unit time the bound state $|\alpha\rangle$ of energy ε_α of a given empty dot, we use the Fermi golden rule,

$$n_\alpha = \frac{2\pi}{\hbar} \sum_k f_D(\varepsilon_k) |M_{k\to\alpha}|^2 \; \delta[E_{WL} + \varepsilon_k - \varepsilon_\alpha - \hbar\omega_{LO}]$$

$$|M_{k\to\alpha}|^2 \equiv \sum_{\vec{Q}} \left| \left\langle k \, ; N_{\vec{Q}} \, \middle| \, H_{e-ph} \, \middle| \alpha; (N+1)_{\vec{Q}} \right\rangle \right|^2. \tag{2.1}$$

We assume a thermal distribution $f_D(\varepsilon_k)$ of free carriers, where k labels the states of the WL continuum and $N_{\vec{Q}}$ the number of LO phonons of wavevector $\vec{Q}$. We assume the LO phonons are monochromatic (energy $\hbar\omega_{LO}$ independent of the phonon

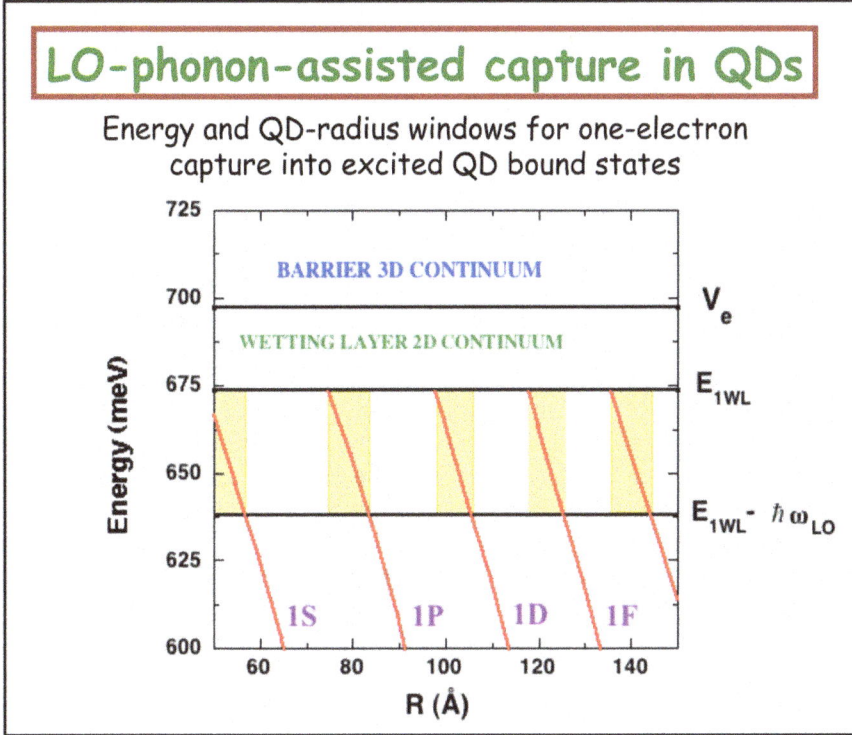

Figure 2.3. Zoom of the high-energy dot states in figure 1.4 near the 2D and 3D edges, to illustrate the energy and dot radius restrictions for the resonant capture into the marginally bound states ψ_{1m} with emission of one LO phonon.

wavevector). The transition towards $|\alpha\rangle$ is thus possible only for carriers with excess kinetic energy $\varepsilon_k = \varepsilon_a + \hbar\omega_{LO} - E_{WL}$. As a consequence, only marginally bound states are accessible, i.e. $0 \leqslant E_{WL} - \varepsilon_a \leqslant \hbar\omega_{LO}$ (see figure 2.3); and the capture rate is proportional to the density of carriers for which an emission is possible, i.e. $n_a \propto f_D(\varepsilon_k = \varepsilon_a + \hbar\omega_{LO} - E_{WL})$. We discuss these two aspects in the following section.

2.1.2 Effects of temperature and gas density

It is worth pointing out that the existence of a well-defined distribution $f_D(\varepsilon_k)$ of free carriers is one important assumption introduced in the calculations, and deserves some comments. It is well accepted that different interactions (electron–electron, electron–phonons) very effectively realize a distribution of photoinjected carriers within the continuum states of the system (WL, barrier). A stationary distribution can be maintained in continuous wave experiments, even if modified by the presence of the dots as compared to those of bare barriers or quantum wells, but this will not be the case in time-resolved experiments (the ones actually performed to access the capture and energy relaxation rates). In the presence of carrier captures by the dots of the ensemble, the gas loses carriers and cools down (it loses the energies of the captured carriers). However, one can consider that these changes have negligible effects on the free carrier distribution characteristics whenever: (i) the gas density is much higher than that of carriers captured by the QDs; (ii) the total energy lost in the captures towards the QDs of the ensemble is small with respect to its total energy, and that a redistribution of the energy deficit occurs amongst the remaining electrons in a very short time, i.e. much shorter than the characteristic capture time; and (iii) the mean capture time is much smaller than the decay time of the free carrier distribution by radiative and/or non-radiative processes independent of the QDs. In this case, one can consider that the free carrier distribution is both time-independent and not affected by the capture. Moreover, the assumption of a capture-independent stationary distribution inhibits the inverse detrapping process. Under such circumstances, the ensemble of free carriers plays the role of a twofold reservoir, at the same time providing particles for LO capture, and ensuring both the energy conservation (altogether with the energy of the emitted phonon) and the irreversibility of the process (without being affected by it). The calculated rates discussed below would thus apply to time-resolved measurements that realize such assumptions; otherwise, the capture becomes a dynamical and possibly nonlinear exchange process involving the QDs and free carrier populations.

Let us assume a 2D gas in the WL, with thermal gas distribution $f_D(\varepsilon_k)$ given by the Fermi–Dirac distribution (accounting for the twofold spin degeneracy),

$$f_D(\varepsilon_k) = 2_{\text{spin}} \left[\exp\left(\frac{E_{\text{WL}} + \varepsilon_k - \mu}{k_B T} \right) + 1 \right]^{-1}. \tag{2.2}$$

The chemical potential μ is an implicit function of the electron areal density n_S,

$$n_S = \frac{1}{S} \sum_k f_D(\varepsilon_k) \rightarrow \frac{1}{2\pi^2} \int_0^\infty d^2k \left[\exp\left(\frac{E_{\text{WL}} + \varepsilon_k - \mu}{k_B T} \right) + 1 \right]^{-1}, \tag{2.3}$$

where the sum has been replaced by the integral in k-space by using

$$\sum_k (\ldots) \rightarrow \frac{S}{(2\pi)^2} \int d^2k \, (\ldots). \tag{2.4}$$

For the free energy dispersion $\varepsilon_k = \hbar^2 k^2 / 2m_*$, the integral can be exactly solved by using the transformation

$$x = \exp\left(\frac{\varepsilon_k}{k_B T} \right) \Rightarrow \frac{dx}{x} = \frac{\hbar^2 k}{m_* k_B T} dk, \tag{2.5a}$$

and

$$\int_1^\infty \frac{dx}{x} \frac{1}{xe^{-a} + 1} = \int_1^\infty dx \left[\frac{1}{x} - \frac{e^{-a}}{xe^{-a} + 1} \right] = \ln(e^a + 1). \tag{2.5b}$$

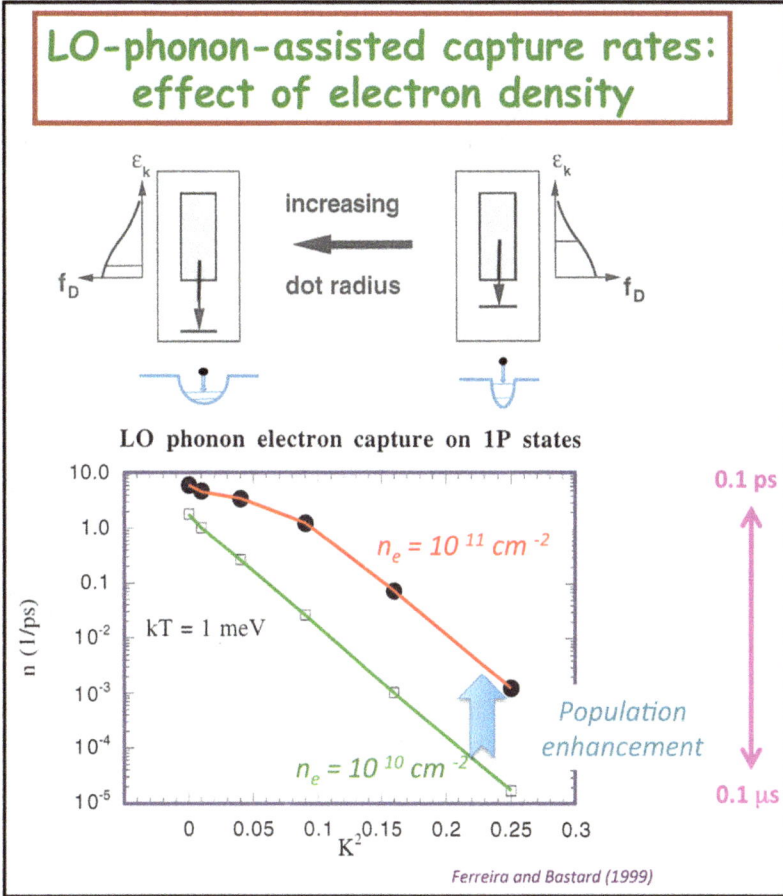

Figure 2.4. Low-temperature capture rate into one 1P state of a cone-like empty QD, for two different electron gas concentrations in the WL. Adapted from Ferreira and Bastard (1999).

Finally, there is

$$\mu - E_{\text{WL}} = k_{\text{B}}T \ln\left[\exp\left(\frac{n_S \pi \hbar^2}{m^* k_{\text{B}} T}\right) - 1\right] \xrightarrow[n_S/T \to 0]{} k_{\text{B}}T \ln\left[\frac{n_S \pi \hbar^2}{m^* k_{\text{B}} T}\right] \ll 0. \quad (2.6)$$

In the non-degenerate limit ($n_S/T \to 0$ and $\mu - E_{\text{WL}} \ll 0$) one obtains

$$f_D(\varepsilon_k) \xrightarrow[n_S/T \to 0]{} 2_{\text{spin}} \exp\left(\frac{-\varepsilon_k + \mu - E_{\text{WL}}}{k_{\text{B}} T}\right) \approx 2_{\text{spin}} \frac{n_S \pi \hbar^2}{m^* k_{\text{B}} T} \exp\left(\frac{-\varepsilon_k}{k_{\text{B}} T}\right), \quad (2.7)$$

which corresponds to the Boltzmann distribution. Finally, in this last limit the scattering rate decreases exponentially with increasing thermal ratio $\varepsilon_k/(k_{\text{B}}T)$, and increases proportionally to the areal gas density.

Figure 2.4 shows the calculated low-temperature rate for capture of one electron into one 1P state of a cone-like dot, for two different densities n_e of the electron gas in the WL, as a function of the dot radius. The dimensionless parameter in the figure is $K^2 = 2m^*\beta_{1P}^2\varepsilon_k/\hbar^2$, where β_{1P} is the in-plane variational extension of the 1P state (see equation (1.16) and figure 1.5). When ε_k increases, the dot radius decreases from its maximum value for a resonant capture from the WL edge ($\varepsilon_k = 0$) to its minimum value when the P state approaches the WL edge ($\varepsilon_k \to \hbar\omega_{LO}$). The order of magnitude of the capture rates can be obtained within a simple model, as discussed in the next section. Let us here make a qualitative analysis of the results in figure 2.4. The calculated rates decrease very fast when ε_k increases, and increase significantly with increasing gas density. To qualitatively explain these trends, one has to consider two factors. First, the dependence with the gas distribution $n_{1P} \propto f_D(\varepsilon_k = \varepsilon_{1P} + \hbar\omega_{LO} - E_{WL})$, as depicted in figure 2.4, means increasing ε_k amounts to decreasing the number of electron available for capture. The second comes from the matrix elements of the Fröhlich potential: in fact, the matrix element in equation (2.1) contains the Fourier transform of the localized state, which decreases when the energy of the initial state increases since its wavefunction oscillates faster and faster. However, for GaAs one has: $\varepsilon_{k=\pi/R} \approx 54$ meV $> \hbar\omega_{LO}$ and the first (population) factor should largely dominate at low densities and temperatures.

It is interesting to stress two important features of these calculations: first, the huge variation (many orders of magnitude) of the characteristic capture time with the QD geometry; and second, the very small radius interval spanned by the K parameter: $K = 0(K^2 = 0.89)$ corresponds to $R \approx 75\text{Å}(\approx 83\text{Å})$ for the QD parameters used in the calculations (see also figure 2.3). These low-temperature results suggest the existence of a strong selectivity for the capture process in a QD ensemble: only the dots for which a quasi-resonant capture is possible can efficiently capture one free carrier.

Let us consider the effect of temperature on capture. Figure 2.5 shows the capture rate into P states for three different temperatures of the WL gas, as a function of the dot radius (same K parameter as in the previous figure) and for $n_S = 10^{10}$ cm^{-2}. Like in the previous case of figure 2.4, the capture rate steeply decreases when the in-plane kinetic energy ε_k increases. The temperature dependence is however more complex. In fact, when the temperature increases the population of low-energy electrons ($\varepsilon_k \approx 0$) decreases slightly, and so too the capture rate. The opposite behaviour occurs for states with energies far from the WL edge ($0 \ll \varepsilon_k \leqslant \hbar\omega_{LO}$), which become more populated at higher temperatures. The size and temperature trends obtained in figures 2.4 and 2.5 are pictured in figure 2.6.

In summary, one finds that an increase of the gas density and/or temperature: considerably activates the non-resonant capture (i.e. of high-energy electrons) while the resonant processes (when $\varepsilon_k \approx 0$) are relatively much less affected; and strongly attenuates the previously mentioned pseudo-selectivity of the capture process at low temperatures and densities, by allowing a more uniform population of different dots of an inhomogeneous ensemble.

2.1.3 Simple analytical model for the capture

The order of magnitude of the calculated capture rates can be interpreted in a simple model, developed in the following. To this end, we consider the matrix elements of

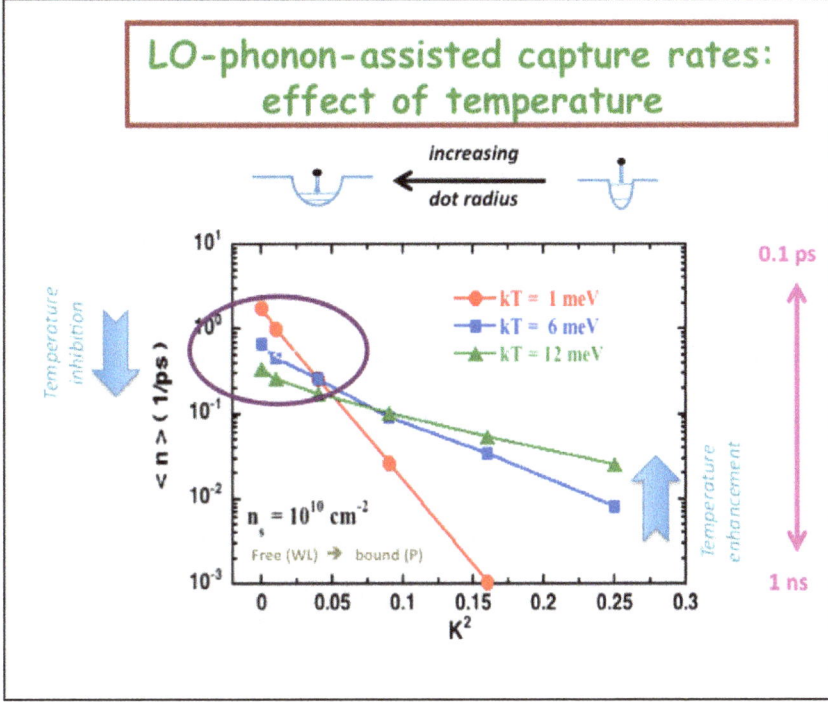

Figure 2.5. Capture rates into one 1P state of a cone-shaped empty QD, for three different temperatures of the WL gas of free electrons. Adapted from Ferreira and Bastard (1999).

the Fröhlich coupling. Taking the WL states as plane waves orthogonalized to the final bound state (see equation (1.17)) we have

$$\psi_{\vec{k}}(\vec{r}) = e^{i\vec{k}\cdot\vec{\rho}}\chi_{WL}(z)/\sqrt{S}$$
$$\tilde{\psi}_{\vec{k}}(\vec{r}) \approx \psi_{\vec{k}}(\vec{r}) - \langle\psi_\alpha(\vec{r})|\psi_{\vec{k}}(\vec{r})\rangle\psi_\alpha(\vec{r}), \tag{2.8}$$

and one has from equation (1.29),

$$\langle\tilde{\psi}_{\vec{k}}(\vec{r}); N_{\vec{Q}}|\,H_{\text{e-ph}}\,|\psi_\alpha(\vec{r}); (N+1)_{\vec{Q}}\rangle \approx \sqrt{\bar{N}_{LO}+1}\,V_{\vec{Q}}\,\langle\tilde{\psi}_{\vec{k}}|\,e^{i\vec{Q}\cdot\vec{r}}\,|\psi_\alpha\rangle \tag{2.9a}$$

$$\Rightarrow |M_{\vec{k}\to\alpha}|^2 = (\bar{N}_{LO}+1)\sum_{\vec{Q}}\left|V_{\vec{Q}}\langle\tilde{\psi}_{\vec{k}}|\,e^{i\vec{Q}\cdot\vec{r}}\,|\psi_\alpha\rangle\right|^2 \tag{2.9b}$$

$$\langle\tilde{\psi}_{\vec{k}}|\,e^{i\vec{Q}\cdot\vec{r}}\,|\psi_\alpha\rangle \approx \langle\psi_{\vec{k}}|\,e^{i\vec{Q}\cdot\vec{r}}\,|\psi_\alpha\rangle - \langle\psi_{\vec{k}}(\vec{r})|\psi_\alpha(\vec{r})\rangle\,\langle\psi_\alpha|\,e^{i\vec{Q}\cdot\vec{r}}\,|\psi_\alpha\rangle, \tag{2.9c}$$

where $\bar{N}_{LO}$ is the temperature-dependent population of LO phonons (Bose distribution). Using the separable wavefunction $\psi_\alpha(\vec{r}) = g_\alpha(z)f_\alpha(\vec{\rho})$ for the bound state (see equation (1.6)) allows the 3D integrals to be split into products of z and radial contributions. The latter are 2D Fourier transforms of $f_\alpha(\vec{\rho})$ or $f_\alpha^2(\vec{\rho})$, which can be

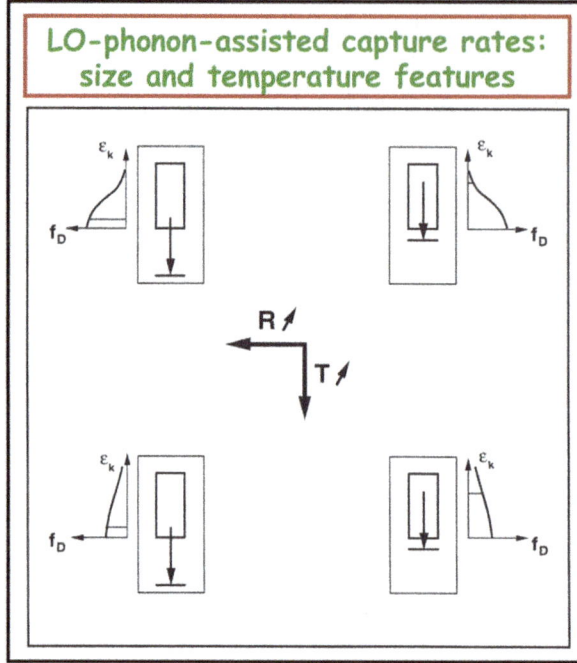

Figure 2.6. Schematic variations of the radius and temperature dependencies of the capture of one electron of the WL gas with emission of one LO phonon.

readily evaluated using the Gaussian variational trial equation (1.16). We shall not develop the calculations here, but only consider the particular case of resonant capture, namely when the bound level is around one-LO phonon below the continuum edge, so that the kinetic energy of the initial carrier is very small: $\varepsilon_a \approx E_{WL} - \hbar w_{LO}$; $\varepsilon_k \approx 0$ ($K \approx 0$ in figures 2.4 and 2.5). In this case, the contribution to the matrix element coming from the orthogonalization to the bound state vanishes (since $\langle \exp\{i\vec{k} \cdot \vec{\rho}\} | f_a(\vec{\rho}) \rangle \xrightarrow{k \to 0} 0$ when $m \neq 0$). Moreover, the z-contribution is approximated by $\langle \chi_{WL}(z) | \exp\{iQ_z z\} | g_a(z) \rangle \approx \langle \chi_{WL} | g_a \rangle$: indeed, since $g_a(z)$ is highly localized, its Fourier transform varies slowly with Q_z for values of Q_z that contribute the most in the calculation. The sum over Q_z can thus be performed,

$$\sum_{Q_z} |V_{\vec{Q}}|^2 \to \frac{L_z}{2\pi} \int dQ_z |V_{\vec{Q}}|^2 = \frac{|C_F|^2}{\Omega_{cr}} \frac{L_z}{2\pi} \int \frac{dQ_z}{Q_z^2 + Q_\perp^2} = \frac{|C_F^2|}{\Omega_{cr}} \frac{L_z}{2\pi} \frac{\pi}{Q_\perp}, \quad (2.10)$$

where Ω_{cr} is the crystal quantizing volume ($\Omega_{cr} = L_z S$) and C_F is the strength of the Fröhlich coupling (see equation (1.29)). Using the form from equation (1.16) for $f_{a=1m}(\vec{\rho})$ one has

$$\int d\vec{\rho} \, e^{i\vec{Q}_\perp \vec{\rho}} f_{1m}(\vec{\rho}) r = 2\beta_{1m} \sqrt{\frac{\pi}{|m|!}} (\beta_{1m} Q_\perp)^{|m|} \exp\left[-(\beta_{1m} Q_\perp)^2/2\right], \quad (2.11)$$

and the sum over $Q_\perp$ can be finally performed,

$$\sum_{\vec{Q}}(\ldots) \rightarrow \frac{S}{(2\pi)^2}\int d\vec{Q}_\perp (\ldots)$$

$$\int d\vec{Q}_\perp \frac{\pi}{Q_\perp} \left| \int d\vec{\rho}\ e^{i\vec{Q}\cdot\vec{\rho}} f_{1m}(\vec{\rho}) \right|^2 = (2\pi)^3 \beta_{1m}\ I_{|m|}$$

$$I_{|m|} = \frac{1}{|m|!}\int dx\ x^{2|m|} e^{-x^2}, \tag{2.12}$$

Collecting the various expressions, one obtains for the resonant capture,

$$n_{1P})_{\text{res}} \approx \left(1 + \bar{N}_{\text{LO}}\right)\frac{m^*}{\hbar^3} f_D(0)\beta_{1m}\ C_F^2\ I_{|m|} \left\langle \chi_{\text{WL}}\big| g_{1m}\right\rangle^2, \tag{2.13}$$

and in the non-degenerate limit,

$$n_{1P})_{\text{res(Boltz)}} \approx 2_{\text{spin}}\left(1 + \bar{N}_{\text{LO}}\right)\frac{\pi\ n_S}{\hbar\ k_B T}\ \beta_{1m}\ C_F^2\ I_{|m|} \left\langle \chi_{\text{WL}}\big| g_{1m}\right\rangle^2, \tag{2.14}$$

The last parameter is the Oz overlap $\langle\chi_{\text{WL}}|g_{1m}\rangle$. Since the WL is very thin, one can roughly approximate its confining potential by $V_{\text{WL}}(z) = -a_{\text{WL}}\ \delta(z)$. The solution for this 1D delta-like potential is readily found to be

$$V_{\text{WL}}(z) = -a_{\text{WL}}\delta(z)$$

$$\Rightarrow \chi_{\text{WL}}(z) = \kappa^{1/2}\exp(-\kappa\ |z|), \quad E_{\text{WL}} = -\hbar^2\kappa^2/2m^*, \tag{2.15}$$

with $\kappa = m^* a_{\text{WL}}/\hbar^2$ in order to fit the derivative discontinuity of $\chi_{\text{WL}}(z)$ at $z = 0$. Using the Gaussian trial function for the z-motion in the QD (see equation (1.3)) one easily obtains the overlap in terms of the error function Φ,

$$\left\langle \chi_{\text{WL}}^n\big| g_{1P}\right\rangle = \kappa^{(n-1)/2}\sqrt{c\sqrt{\pi}/2}\ e^{-a^2/2}\left\{e^{\gamma_-^2}\left[1 - \Phi(\gamma_-)\right] + e^{\gamma_+^2}\left[1 - \Phi(\gamma_+)\right]\right\}$$

$$\gamma_\pm = (nc \pm a)/\sqrt{2}, \quad a = z_{1m}/\xi_{1m}, \quad c = = \kappa\xi_{1m}. \tag{2.16}$$

By imposing $E_{\text{WL}} \approx -21$ meV (see figure 1.4), one finds $1/\kappa \approx 51$ Å, and for $n = 1$, $\xi_{1P} \approx 14$ Å and $z_{1P} \approx 5$ Å (see figure 1.5), one finally obtains $\langle\chi_{\text{WL}}|g_{1P}\rangle \approx 0.8$.

Using $C_F \approx 196$ meV $\text{Å}^{1/2}$ (for GaAs; see section 1.5.6), $I_1 = \pi^{1/2}/4$, $\beta_{1P} \approx 35$ Å from figure 1.5, $k_B T = 1$ meV, $\langle\chi_{\text{WL}}|g_{1P}\rangle \approx 0.8$ and $n_S = 10^{10}$ cm^{-2}, one has from equation (2.13) $n_{1P} \approx 3.0$ ps^{-1}. The agreement with numerical calculations for $K^2 = 0$ in figure 2.4 is good owing to the approximations made. For a denser gas, the Boltzmann approximation is no longer appropriate: one has for $n_S = 10^{11}$ cm^{-2} and $k_B T = 1$ meV that $f_D(0)/2 \approx 3.4$ using equation (2.7), which is unphysical. Using instead $f_D(0) \approx 2$ in equation (2.13), one obtains $n_{1P} \approx 10.0$ ps^{-1} for $n_S = 10^{11}$ cm^{-2}, also in reasonably good agreement with the $K^2 = 0$ value in figure 2.4.

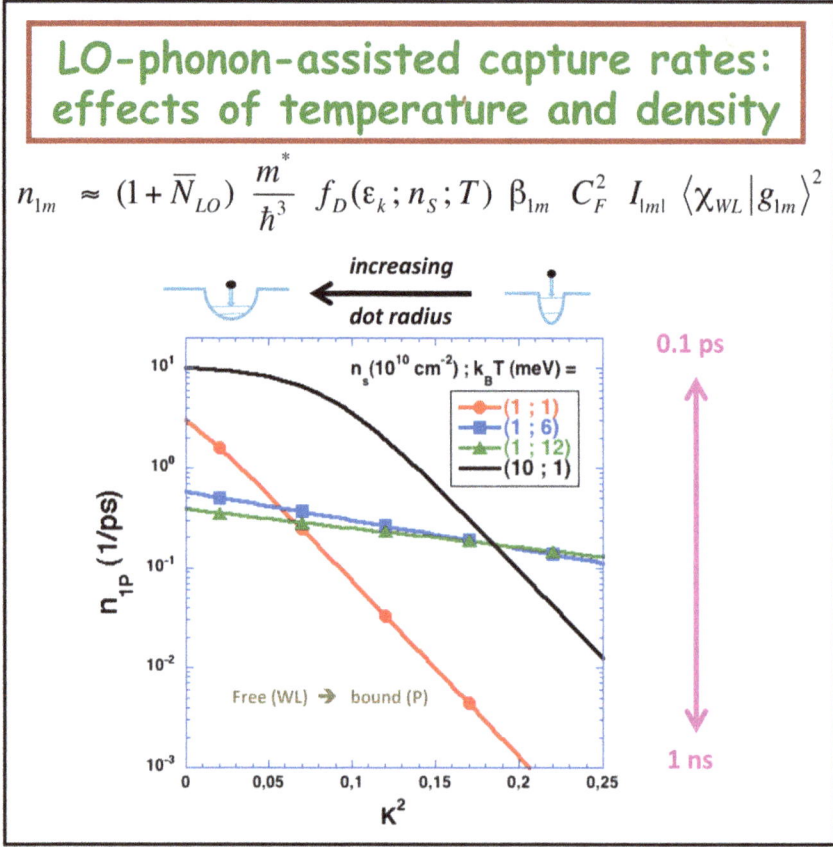

Figure 2.7. Capture rate into one 1P state of a cone-shaped empty QD, for different temperatures and densities of the WL gas of free electrons, obtained with equation (2.17; see text). To be compared to the results in figures 2.4 and 2.5.

The role of the terms neglected in the last calculations can be seen if we assume that the previous formula (2.13) applies also in the non-resonant case,

$$n_{1P} \approx \left(1 + \bar{N}_{LO}\right)\frac{m^*}{\hbar^3} f_D(\varepsilon_k; n_S; T)\beta_{1m} C_F^2 I_{|m|} \left\langle \chi_{WL}|g_{1m}\right\rangle^2, \qquad (2.17)$$

with $f_D(\varepsilon_k = \varepsilon_a + \hbar\omega_{LO} - E_{WL})$ radius-dependent. We show in figure 2.7 the results of calculations for different temperatures and densities (same parameters as in the two previous figures) using the last formula equation (2.17). This last approximation underestimates the non-resonant capture rate, as can be seen from a direct comparison of the results in figures 2.4, 2.5 and 2.7. It nonetheless clearly points out the important gas characteristic (temperature and density) dependences on the electron capture.

2.2 Phonon-assisted capture by a charged dot

Let us consider now the capture assisted by the emission of one LO phonon in the case of a charged dot, i.e. already hosting a bound carrier. In fact, as mentioned in

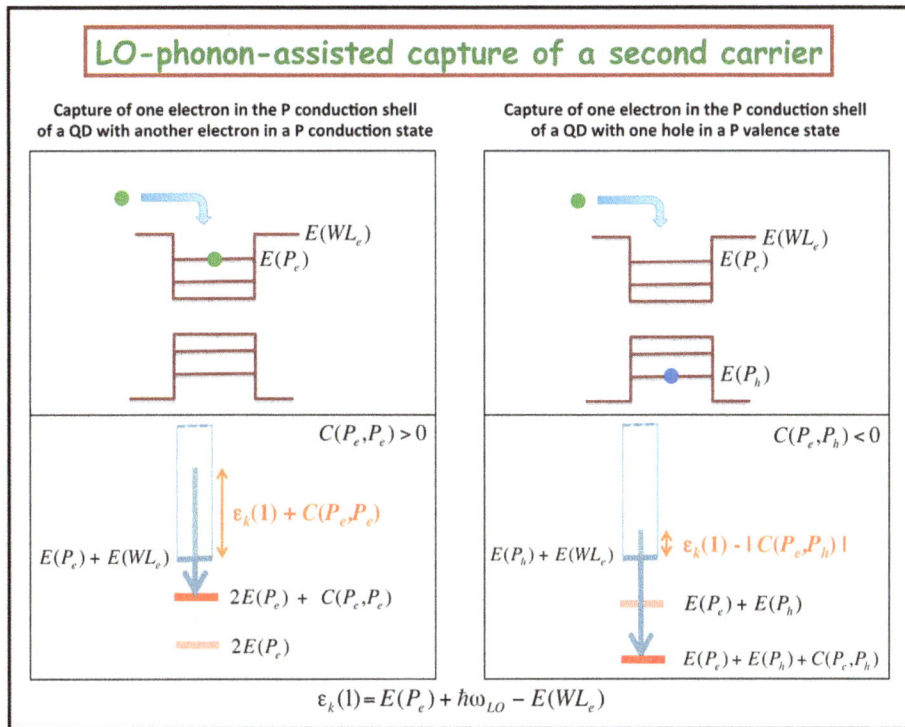

Figure 2.8. Schematic picture (upper) and energy (lower) representations of the capture of one electron by a QD hosting another electron (left) or a hole (right) in its P level. See text.

section 1.4, a QD can bind various charges. Such a multi-particle state may result from various consecutive captures. It is thus interesting to ascertain to what extent the presence of a first carrier affects the capture of a second one.

We show schematically in figure 2.8 the capture of one electron in the P shell of a dot charged with either another electron or a hole in its corresponding P shell. The main difference with the first capture is the existence of a final-state interaction between the carriers. In fact, the Coulomb interaction in the initial state involves one localized and one delocalized carrier and can be neglected, whereas in the final state both the electron–electron $C(P_e, P_e) > 0$ and electron–hole $C(P_e, P_h) < 0$ Coulombic energies are important (a few tens of mili-electron volts; see, for example, figure 1.7). As a consequence, the initial kinetic energy of the second carrier decreases (increases) as compared with the value $\varepsilon_k(1)$ for capture of the first electron, due to the attractive (repulsive) interaction with the initially bound carrier of opposite (same) sign, as shown in the lower panel of figure 2.8.

2.2.1 Two opposite charges

Figure 2.9 shows the calculated capture rates (involving only the P shells) when two opposite charges are involved: capture of one electron in a positively charged dot (left) or the contrary (right).

The dot radius intervals in the graphics are in each case the same as the ones where capture of the second carrier alone by the empty dot is possible (i.e. without Coulomb interaction and the corresponding rates are given by the empty circles). The calculations were done for a density of 10^{11} cm^{-2} free particles (either electrons or holes) and for $T = 300$ K. It is worth noting three main features:

- In 'large' dots, the capture of a second carrier is completely inhibited since the important electron–hole interaction $C(P_e, P_h) < 0$ renders the final state too low in energy to be reached by a single LO phonon emission: the **Coulomb blockade** effect.
- When the capture is energetically possible, its rate is reinforced essentially because its initial kinetic energy is decreased by the attractive final-state interaction (see right-lower panel in figure 2.8): the **Coulomb enhancement** effect. For a non-degenerate gas, the enhancement factor (ratio between the second and the first capture rates) is $n_P(2)/n_P(1) = \exp[|C(P_e, P_h)|/k_B T]$. In figure 2.9 this factor is weak since $|C(P_e, P_h)| \approx k_B T$, but it is of paramount importance at low temperatures.
- The curves for the rates with Coulomb corrections appear similar to the one without correction shifted towards smaller radius. This is a consequence of the fact that the main effect of the first bound carrier is to shift downwards

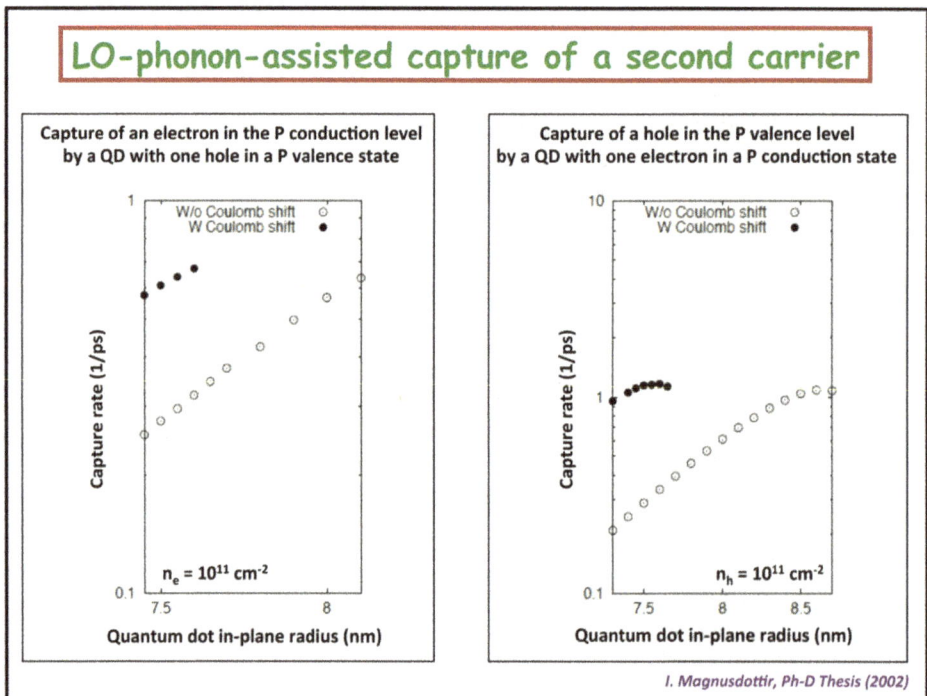

Figure 2.9. Rates for the capture of a second carrier of opposite charge to the bound one, with (dark circles) or without (open circles) final-state Coulomb interaction. Only P states of a cone-like QD are involved. $T = 300$ K. From Magnusdottir (2002).

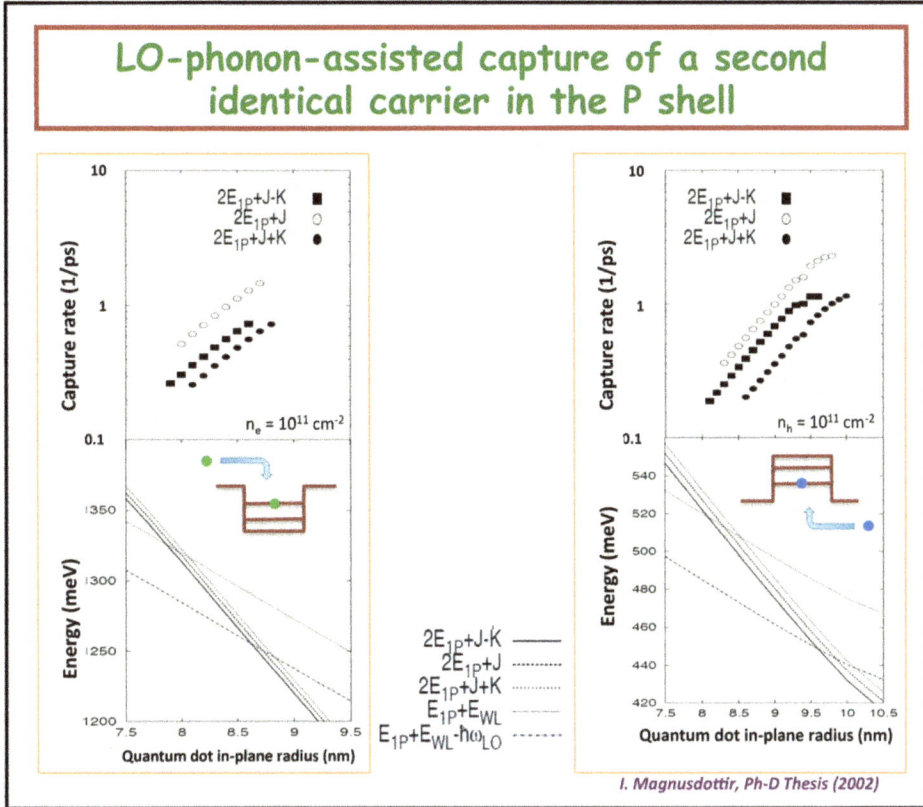

Figure 2.10. Rates (upper panels) and different two-electron energies (lower panels) for the capture of a second carrier of the same charge to the bound one, with final-state Coulomb direct and exchange interactions: two electrons (left) or two holes (right) processes. Only P states of a cone-like QD are involved. Lower panel insets: pictures of the capture processes. Adapted from Magnusdottir (2002).

the final state energy for the second capture, which can be envisioned as an apparent radius shift: a lower final state appears as the final state of a 'bigger' dot.

A more accurate calculation would also account for the wavefunction modification induced by the bound carrier, especially for the 'free' initial carrier (because of the long-range Coulomb interaction); such a calculation has not, to our knowledge, been performed.

2.2.2 Two identical charges

Figure 2.10 presents the rates (upper panels) and two-electron energies (lower panels) for the capture of a second carrier by a dot hosting a carrier of the same charge, as a function of its radius (only P states are considered). The figures on the left (right) are for two electrons (two holes).

The main difference with the previous case is that identical particle states should comply with Pauli's principle. Anti-symmetrization brings negligible corrections to the initial state energies. The case of two-electron states in the P shell (here, the final state for the capture) was discussed in section 1.4. The same analysis applies for two-hole states. In figure 2.10, the lower panels represent the energies $2E_{1P} + C_{pp}$ of the three levels (two singlet ($C_{pp} = J + K$ or J) and one triplet ($C_{pp} = J - K$)) for the dot hosting two indistinguishable carriers in its P shell, as a function of its radius. The other two energies in these panels fix the radius interval where the second capture is possible,

$$E_{1P} + E_{WL} \geqslant 2E_{1P} + C_{pp} \geqslant E_{1P} + E_{WL} - \hbar\omega_{LO}. \quad (2.18)$$

The first inequality represents the condition for energetic stability of the final state: in fact, the 'small' dots, for which the P shell is too near the WL continuum, cannot bind two identical carriers in this shell (when the inequality is not fulfilled, the final state becomes a resonance).

The second inequality simply represents the possibility of finding a free initial state such that the total energy is conserved (final state not too deep compared to $\hbar\omega_{LO}$). For each singlet or triplet state, the second capture is possible when the two inequalities are fulfilled (in between the two curves with weaker slopes). The extreme limits are represented in the schemes in figure 2.11.

Concerning the calculated rates, one should notice three main features:

- In 'small' dots, the capture of a second identical carrier is completely inhibited, since, as already mentioned, the important Coulomb interaction $C_{pp} > 0$ renders the final state unstable: the **Coulomb suppression** effect.

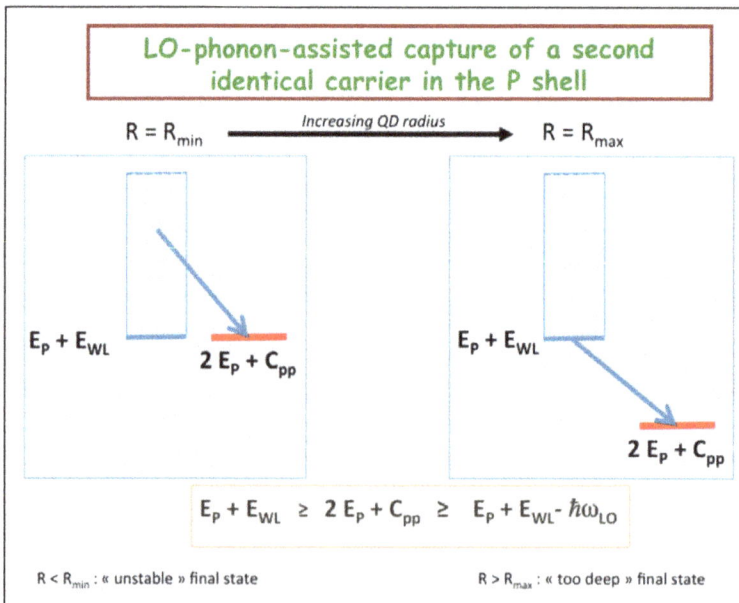

Figure 2.11. Scheme of two-electron energies illustrating the existence of a radius dot interval wherein a second capture is possible.

- When possible, capture of the second carrier is less favourable essentially because its initial kinetic energy is increased by the repulsive final-state interaction: the **Coulomb inhibition** effect. For a non-degenerate gas, the blocking factor (ratio between the second and first capture rates) is $n_P(2)/n_P(1) = \exp[-\mathbf{C_{pp}}|/k_B T]$. In figure 2.10 this factor is weak since $\mathbf{C_{pp}} \approx k_B T$, but it is of paramount importance at low temperatures.
- The curves for the rates with Coulomb corrections appear similar to the ones without correction (empty circles in figure 2.9) shifted towards bigger radius. This is a consequence of the fact that the main effect of the first bound carrier is to blue-shift the final state energy for the second capture, which can be envisioned as an apparent radius shift: a final state with larger energy appears as the final state of a smaller dot.

In summary, the capture of a second carrier may display different kinds of Coulomb-related effects depending on the signs of the first and second charges and on the dot radius: blocking, suppression, enhancement or inhibition. These reflect the strong final-state interaction for either identical (repulsive interaction) or distinguishable (attractive interaction) carriers.

More generally, capture assisted by the emission of one optical phonon, when allowed, is an important mechanism to transfer free charges into the dots.

2.3 Coulombic-assisted capture by one empty dot

Let us consider now the trapping processes assisted by Coulombic interactions among the free carriers. It is well known that in bulk semiconductors and quantum wells these interactions very effectively redistribute the total kinetic energy of an ensemble of electrons among themselves, so as to generate or maintain a (possibly hot) Fermi–Dirac distribution. In the presence of the dots, the interactions among free carriers also assist the capture process, since one free electron can be scattered towards a bound state of the dot after a Coulombic exchange with another free carrier (of identical charge or not), the latter being promoted in the end to a more energetic free state. The gas ends up hotter (it gains binding energy) and has fewer carriers. However, one can consider that these are negligible effects on the free carrier distribution characteristics whenever: its density is much higher than the one of bound carriers; and the energy it gains in the capture is small with respect to its total energy, and can be redistributed among the ensemble of free electrons in a very short time (shorter than the capture time, so as to inhibit the inverse detrapping process). In this case, one can consider that the free carrier distribution is not affected by the capture. The ensemble of free carriers thus plays the role of a twofold reservoir, at the same time providing particles for the capture, and ensuring both the energy conservation and the irreversibility of the process (without being affected by it). This Coulomb-assisted process is usually termed Auger after the well-known original effect in the physics of many-electrons atoms, when the filling of an inner-shell vacancy by an electron of an excited bound orbit is accompanied by the emission of another electron from a different orbit of the same atom. This analogy will be used in the next chapter, where we shall consider the Coulomb-mediated intra-dot relaxation of a doubly charged dot, whereas in the present paragraph the dot is initially empty and the two interacting electrons are in the continuum. Note also that, as we shall see in this and the next chapter, in QDs both electrons and holes may participate in Auger-like processes.

2.3.1 Scattering rate and indistinguishability effects

The capture rate into the discrete state $|d\rangle$ is given by the general expression

$$R_d = \frac{2\pi}{\hbar} \sum_{\substack{\text{spin} \\ \text{config}}} \sum_{\vec{k}_C, \vec{k}_S, \vec{k}_F} \left| \left\langle \vec{k}_C, \vec{k}_S \right| V_C \left| d, \vec{k}_F \right\rangle \right|^2 F_{\vec{k}_C, \vec{k}_S, \vec{k}_F} \, \delta \left[E_f - E_i \right]$$

$$F_{\vec{k}_C, \vec{k}_S, \vec{k}_F} = f_{\text{FD}}\left(E\left(\vec{k}_C\right)\right) f_{\text{FD}}\left(E\left(\vec{k}_S\right)\right) \left[1 - f_{\text{FD}}\left(E\left(\vec{k}_F\right)\right)\right]. \tag{2.19}$$

Two carriers in states $|\vec{k}_S\rangle$ and $|\vec{k}_C\rangle$ interact via the Coulomb potential V_C and scatter each other, whereby one ends in the discrete state $|d\rangle$ (a bound orbit) while the other is promoted to the high-energy final state $|\vec{k}_F\rangle$. The Dirac delta ensures that total energy is conserved in the process, while the population factors indicate that the rate is proportional to the initial occupations of $|\vec{k}_S\rangle$ and $|\vec{k}_C\rangle$, and to the available empty final states in the continuum (the dot is assumed to be initially empty). Note that the population distributions are Fermi–Dirac ones $f_{\text{FD}}(E) = f_D(E)/2$, with $f_D(E)$ defined in equation (2.2): the latter incorporates spin degeneracy, while here the spin contribution should come from the 'spin config' summation. The carriers involved may be two electrons, two holes or one electron and one hole.

It is important to mention the indistinguishability effects in processes involving identical carriers (only electrons or only holes). In this case, it is of course impossible to distinguish between the captured carrier, the scattered one and all the others. However, owing to the binary nature of the Coulomb interaction $V_c(1, 2)$, the matrix element for scattering depends only upon the involved initial and final states, and contains both direct (J) and exchange (K) terms,

$$\Psi_{n,m,\pm} = \frac{1}{\sqrt{2}} \left\{ \psi_n(1)\psi_m(2) \pm \psi_n(2)\psi_m(1) \right\} \tag{2.20a}$$

$$\Rightarrow \left\langle \Psi_{n,m,\sigma} \right| V_c \left| \Psi_{N,M,\Sigma} \right\rangle = \frac{1 + \sigma\Sigma}{2} J + \frac{\sigma + \Sigma}{2} K = \delta_{\sigma,\Sigma}(J + \sigma K)$$

$$\begin{cases} J = \left\langle \psi_n(1)\psi_m(2) \right| V_c(1, 2) \left| \psi_N(1)\psi_M(2) \right\rangle \\ K = \left\langle \psi_n(1)\psi_m(2) \right| V_c(1, 2) \left| \psi_N(2)\psi_M(1) \right\rangle, \end{cases} \tag{2.20b}$$

where n, m, N and M are the four states involved in the scattering and σ, $\Sigma = +1$ for a symmetric (-1 for an anti-symmetric) two-carrier orbital wavefunction (see also equation (1.19)). The matrix element vanishes for $\sigma = -\Sigma$: in fact, since the Coulomb interaction $V_c(1, 2)$ between carriers '1' and '2' is independent of their spins, the symmetric ($|i\rangle = |\Psi_{n,m,+}\rangle$; $|f\rangle = |\Psi_{N,M,+}\rangle$) or anti-symmetric ($|i\rangle = |\Psi_{n,m,-}\rangle$; $|f\rangle = |\Psi_{N,M,-}\rangle$) nature of the initial ($|i\rangle$) and final ($|f\rangle$) states is preserved during the collision. If one additionally neglects the spin-dependent (exchange) correction on the energy of the electron states (in particular the bound one), then $E_f - E_i$ is also independent of the spin configuration. Thus, considering the existence of only one 'singlet' spin configuration ($\sigma = \Sigma = +1$) and of three 'triplet 'ones ($\sigma = \Sigma = -1$), one gets $\sum_{\text{spin config}} |...|^2 \Rightarrow 3|J - K|^2 + |J + K|^2$.

The previous results hold for $n \neq m$, so that in the summations in equation (2.19) one should exclude $\vec{k}_S = \vec{k}_C$. In this latter case, the initial state is the singlet $\psi_{\vec{k}_S}(1)\psi_{\vec{k}_C}(2)$, the final state is $\Psi_{\vec{k}_F,d,+}$ and the additional singlet contribution is $|J + K|^2/2$, with $J = K$ for $\vec{k}_S = \vec{k}_C$. The total singlet rate is thus

$$R_d^{\text{sing}} = \frac{2\pi}{\hbar} \sum_{\substack{\text{spin} \\ \text{config}}} \sum_{\vec{k}_C, \vec{k}_S, \vec{k}_F} \left| (J + K)_{\vec{k}_F,d}^{\vec{k}_C,\vec{k}_S} \right|^2 F_{\vec{k}_C,\vec{k}_S,\vec{k}_F} \left(1 - \frac{1}{2}\delta_{\vec{k}_C,\vec{k}_S} \right) \delta[E_f - E_i]. \quad (2.21)$$

For an electron–hole Auger process there is instead $K = 0$ and $\sum_{\text{spin config}} |...|^2 \Rightarrow 4|J|^2$ because of the four independent spin configurations $\langle\langle\uparrow,\uparrow\rangle\rangle, \langle\langle\uparrow,\downarrow\rangle\rangle, \langle\langle\downarrow,\uparrow\rangle\rangle$ and $\langle\langle\downarrow,\downarrow\rangle\rangle$ for the distinguishable electron and hole particles.

2.3.2 Gas density dependence of the Auger capture

The full circles in figure 2.12 show the variation of the rate for capture of one electron into one state of the P shell of an empty dot, due to the Coulombic scatterings among free electrons of a high-temperature ($T = 300$ K) thermal distribution in the WL, as a function of the gas density (the process is schematically presented in the inset: Auger e–e). In her calculations using equation (2.19), Magnusdottir (2002) used $\sum_{\text{spin config}} |...|^2 \approx 8|J|^2$. From the results we clearly see that the rate increases quickly (quadratically) with the gas density for low densities: $R = C_{ee} (n_{2D})^2$, with $C_{ee} \approx 5 \times 10^{-20}$ m^{4}s^{-1}. A square-law dependence follows straightforwardly when we consider the expression for the capture rate in equation (2.19) (see also inset of figure 2.12). In the non-degenerate limit the Fermi–Dirac occupancy is proportional to the whole gas density n_{2D} (see equation (2.7)) so that in this limit the product of the two initial population factors in equation (2.19) leads to the quadratic dependence observed in figure 2.12 at low densities.

For the sake of comparison, figure 2.12 also presents the rate for capture assisted by the emission of optical phonons. As previously discussed, this process increases more slowly (linearly) with gas density, and the Auger process becomes more efficient at high electron densities. For instance, for a high-density excitation of 10^{11} cm^{-2} one has $1/R \approx 20$ ps capture time per dot, whereas this time increases very fast to about 2 ns for a moderate optical injection (10 times fewer free carriers). In conclusion, the Auger capture rate is a nonlinear function (a quadratic law) of the injection power, at least at low injections such that the assumption of an initial empty dot applies, whereas phonon-assisted capture is proportional to the carrier density and therefore also the injection power.

In practice, an optical excitation in the dot continuum injects high-energy electrons and holes in the conduction and valence bands. Thus, in general, Auger scattering may lead to the capture of either an electron or a hole, assisted by either

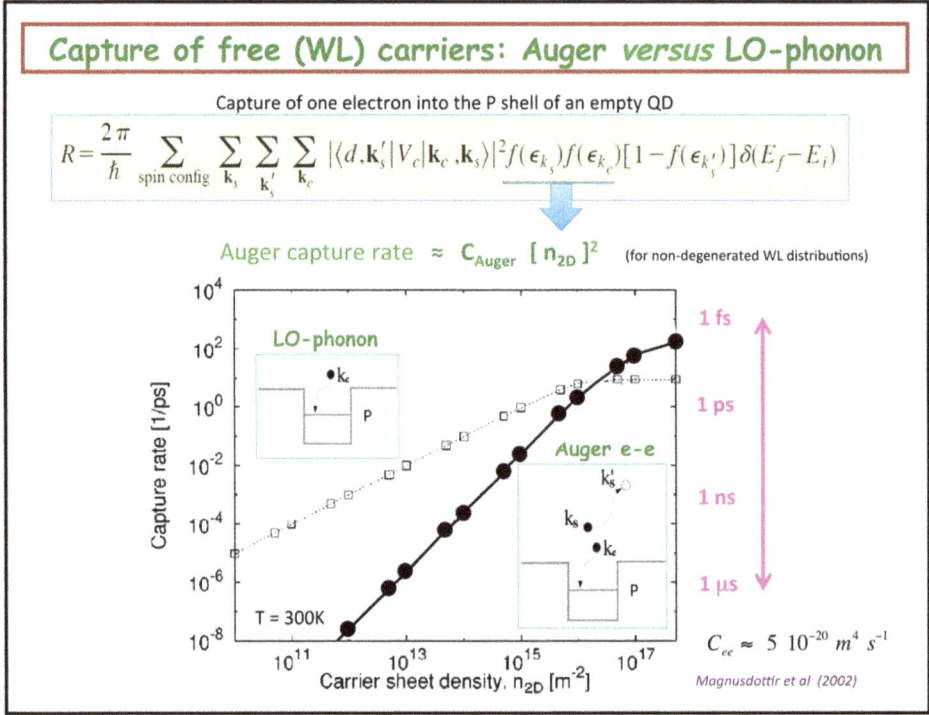

Figure 2.12. Auger capture rate as a function of the free WL gas density (full circles). Capture assisted by emission of one LO phonon is also presented for comparison (empty symbols). $T = 300$ K. Truncated cones with basis angle $30°(= \pi - \alpha)$; $V_b = 697$ meV; $m^* = 0.07m_0$ for electrons and 0.34 for holes; fixed height $h = 30$ Å. Adapted from Magnusdottir *et al* (2003).

the electron or the hole gas. For all cases, the capture rate increases with the square of the free particle gas density in the non-degenerate gas regime. Let us define the Auger coefficients C_{ab} for capture of particle 'a', due to its interaction with a gas of carriers 'b' (a, b = 'e' for electron and 'h' for hole gas), in the regime where the total rate is written $R_{ab} = C_{ab} (n_{2D})^2$. In figure 2.13, the C_{ab} are the Auger coefficients for capture towards the ground S (upper panel) or excited P (lower panel) dot orbital. The dark circle in the lower panel represents the case a = b = 'electron' considered in the previous figure. We see that some dots possess more than one efficient Auger channel, but most of the time only one path dominates. For equivalent electron and hole densities of free carriers, hole capture is in many cases favoured with respect to electron capture. For the dominant channel (Auger h–e) in figure 2.13, note that holes have more bound states and a smaller confining potential and, at the same time, electrons have a smaller effective mass and a correspondingly smaller wave-vector variation $\Delta k = |\vec{k}_F - \vec{k}_S|$, and that both features favour the Coulomb matrix elements. Note also that each Auger coefficient is larger (barely for holes but more clearly for electrons) for capture towards the excited 1P state, for which the amount of relaxed energy is smaller. However, and similarly to the case of capture by emission of optical phonons, a satisfactory interpretation of all the calculated trends is hard to provide. Part of the difficulty comes from the obvious fact that any capture

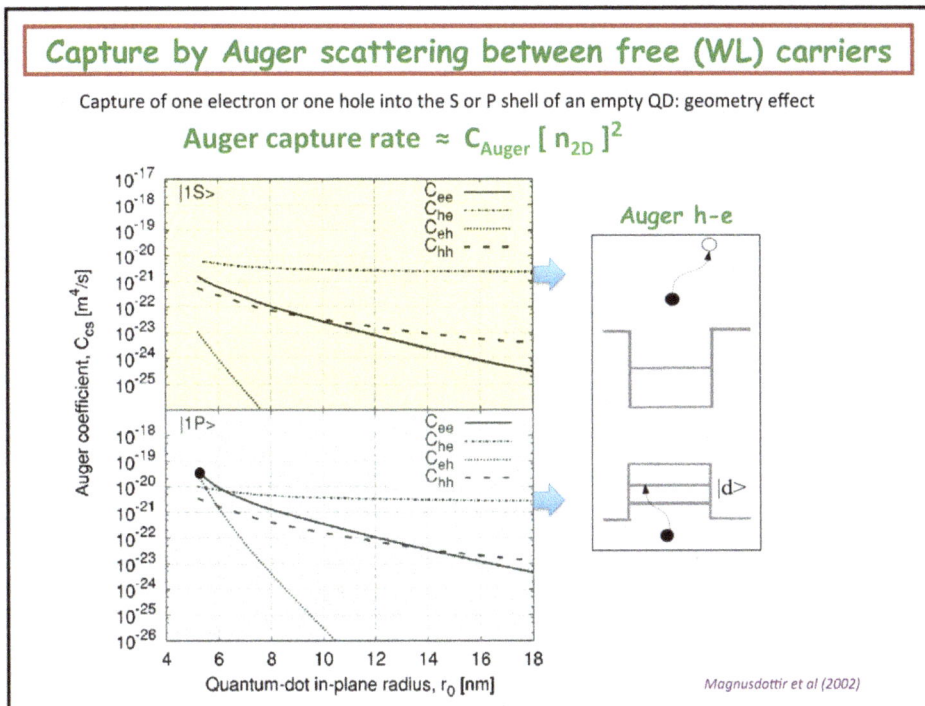

Figure 2.13. Auger coefficients C_{ab} for capture of particle 'a' towards its ground S (upper panel) or excited P (lower panel) dot orbital, due to its interaction with a gas of carriers 'b' (a, b = 'e' for electron and 'h' for hole gas). The dark circle in the lower panel represents the case a = b = electron considered in the previous figure. The arrows indicate the C_{he} coefficients for the process shown on the right scheme. Adapted from Magnusdottir *et al* (2003).

process mixes properties of the free and bound states, but also that the two scattering mechanisms considered in this book involve matrix-elements of a potential with Fourier decomposition that diverges at small wavevectors. However, as we did for the LO-assisted capture, we provide in the following a simplified model for the evaluation of the Auger rate.

2.3.3 A simple model for the Auger capture coefficients

Let us consider the simple situation of a low-density gas, neglect any Coulomb correction to the energies, write $E_{WL}(\vec{k}) = E_{WL} + \hbar^2 \vec{k}^2 / (2m*)$ and define the energy distance between the bound state and the WL edge as $|e_d| \equiv E_{WL} - E_d$. If we furthermore focus on a deep final dot state ($|e_d| \gg \hbar^2 \vec{k}_{C,S}^2 / (2m_{C,S}^*)$), the energies and wavevectors of the initial WL states $|\vec{k}_C\rangle$ and $|\vec{k}_S\rangle$ are small (low-temperature situation), and one can approximate $E_f - E_i \approx E_{WL}(\vec{k}_F) + E_d - 2E_{WL} \equiv \hbar^2 \vec{k}^2 / (2m_S^*) - e_d$.

Using the 2D Fourier decomposition of the Coulomb potential,

$$V_C(1, 2) = \frac{e^2/k_r}{\left[|\vec{\rho}_1 - \vec{\rho}_2|^2 + |z_1 - z_2|^2 \right]^{1/2}} = \frac{2\pi e^2}{k_r S} \sum_{\vec{q}} \frac{e^{i\vec{q} \cdot (\vec{\rho}_1 - \vec{\rho}_2)}}{q} e^{-q|z_1 - z_2|}, \quad (2.22)$$

where $k_r = 4\pi\varepsilon_0\varepsilon_r$, and the separable forms $\Psi_d(z, \vec{\rho}) = \phi_d(z)\psi_d(\vec{\rho})$ and $\Psi_{\vec{k}}(z, \vec{\rho}) = \phi_{WL}(z)e^{i\vec{k}\cdot\vec{\rho}}/\sqrt{S}$ for the wavefunctions of the bound and WL states, the matrix element for the direct-Coulombic interaction can be approximated by

$$J = \langle k_C, k_S | V_C(1, 2) | d, k_F \rangle = \frac{2\pi\,e^2}{k_r\,S}\sum_{\vec{q}} \frac{I_z(q)I_\perp(\vec{q})}{q}$$

$$I_\perp(\vec{q}) \approx \left\langle \frac{1}{\sqrt{S}}\frac{1}{\sqrt{S}} \middle| e^{i\vec{q}\cdot(\vec{\rho}_1 - \vec{\rho}_2)} \middle| \psi_d(\vec{\rho}_1)\frac{e^{i\vec{k}_F\cdot\vec{\rho}_2}}{\sqrt{S}} \right\rangle$$

$$= \delta_{\vec{k}_F - \vec{q}}\frac{1}{\sqrt{S}}\int d\vec{\rho}\; \psi_d(\vec{\rho})e^{i\vec{q}\cdot\vec{\rho}} \equiv \delta_{\vec{k}_F - \vec{q}}\frac{1}{\sqrt{S}}\, \tilde{\psi}_d(\vec{k}_F)$$

$$I_z(q) = \langle \varphi_{WL,C}(z_1)\varphi_{WL,S}(z_2) | e^{-q|z_1 - z_2|} | \phi_d(z_1)\varphi_{WL,S}(z_2) \rangle, \tag{2.23}$$

with $\tilde{\psi}_d(\vec{k}_F)$ the Fourier transform of the in-plane part of the bound-state wavefunction at the wavevector of the excited electron. We can readily convince ourselves that the approximated matrix element for the exchange-Coulombic interaction $\langle k_C, k_S | V_C(1, 2) | k_F, d \rangle$ is equal to the direct one, so that $\sum_{\text{spin config}} |\ldots|^2 \Rightarrow 4\,|J|^2$. Moreover, since the relation between the 2D gas density and the population distribution,

$$\frac{1}{S}\sum_{\vec{k}_C} f_{FD}\big(E(\vec{k}_C)\big) = \frac{1}{S}\sum_{\vec{k}_S} f_{FD}\big(E(\vec{k}_S)\big) = n_{2D}/2, \tag{2.24}$$

one obtains the approximate scattering rate

$$R_d \approx \frac{2\pi}{\hbar}4\frac{n_{2D}^2}{4S}\left(\frac{2\pi e^2}{k_r}\right)^2 \sum_{\vec{q}} \left| \frac{I_z(q)\tilde{\psi}_d(\vec{q})}{q} \right|^2 \delta\left[\hbar^2\vec{q}^2/(2m_S^*) - e_d\right]$$

$$= \frac{n_{2D}^2}{4}\frac{m_S^*}{\hbar}\left(\frac{\pi e^2}{k_r\hbar}\right)^2 \left| \frac{2I_z(q_d)}{q_d} \right|^2 \left\langle \left|\tilde{\psi}_d(\vec{q}_d)\right|^2 \right\rangle, \tag{2.25}$$

where $\langle |\tilde{\psi}_d(\vec{q}_d)|^2 \rangle$ is the angular average $(1/(2\pi))\int d\theta_{\vec{q}}\, |\tilde{\psi}_d(\vec{q}_d)|^2$ and $q_d = \sqrt{2m_S^*e_d/\hbar^2}$ the wavevector of the excited electron. This expression can be further simplified by using the identity $qe^{-q|z_1 - z_2|}/2 \xrightarrow{q\to\infty} \delta[z_1 - z_2]$ to get

$$R_d \approx n_{2D}^2\frac{m_S^*}{\hbar^3}\left(\frac{2\pi e^2}{k_r q_d^2}O_z\right)^2 \left\langle \left|\tilde{\psi}_d(\vec{q}_d)\right|^2 \right\rangle, \tag{2.26}$$

where $O_z = \int_{-\infty}^{+\infty} dz\phi_d(z)\varphi_{WL,C}(z)[\phi_{WL,S}(z)]^2$ is an effective z-overlap between the bound and WL functions. Using the result in equation (2.16) with $n = 3$ and the same parameters for the conduction band WL state, one gets $O_Z \approx 10^{-2}$Å^{-1}. For the variational state $\psi_{1m}(\vec{\rho})$ in equation (1.16) there is

$$\left\langle \left| \tilde{\psi}_{1m}(\vec{q}_{1m}) \right|^2 \right\rangle = \frac{4\pi\beta_{1m}^2}{|m|!} (\beta_{1m}q_{1m})^{2|m|} \exp\left(-\beta_{1m}^2 q_{1m}^2\right)$$

$$q_{1m} = \sqrt{2m_S^* e_{1m}/\hbar^2}. \tag{2.27}$$

So, for the ratio of capture rates towards one state of the S and P shells, we obtain

$$R_{1P}/R_{1S} \approx \left[2m_S^* \beta^2 e_{1P}/\hbar^2 \right] \exp\left(+2m_S^* \beta^2 (E_{1P} - E_{1S})/\hbar^2\right). \tag{2.28}$$

In conclusion, even if this result does not apply near the 1P edge (at $e_{1P} = 0$), it shows that the ratio of the two capture rates strongly depends on the actual dot parameters: for tightly bound states, e_{1P} increases while $E_{1P}-E_{1S}$ decreases with increasing dot radius.

2.3.4 Auger capture of a second carrier

Auger capture of a second carrier has not so far, to the best of our knowledge, been studied. However, similar to the optical phonon case, one can assume that the main effect of the first capture is to modify the two-carrier final-state energy, by the repulsive (attractive) interaction of particles of the same (different) charges. The dot seems to have an 'effective' radius, slightly smaller (larger) for the capture of a second carrier of the same (opposite) charge. As C_{Auger} decreases when the radius increases, the formation of doubly charged dots would be favoured. This counter-intuitive conclusion (because of Coulomb repulsion) is opposite to the one obtained for phonons, which tend to neutralize the dot. By the same argument, one would expect that the capture directly towards the S shell becomes enhanced when the dot has already captured one carrier of same charge in an excited orbit.

2.4 Conclusion

In conclusion to this chapter, free carriers are preferably captured into excited orbits near the continuum edge. The capture rates display sizeable variations with respect to dot parameters, and also strongly depend upon experimental conditions (temperature, density of photoinjected carriers,...). The main difference between Auger-related and phonon-assisted captures is that the Coulomb carrier–carrier interaction may *a priori* relax any amount of energy, in contrast to the monochromatic LO modes, and thus Auger processes do not display strict energy and radius windows for capture (see, for example, figure 2.3). However, the huge variations of some Auger coefficients with dot geometry still suggest some selectivity in Coulomb-assisted capture efficiency inside the QD distribution. In practice, only dots with (at least) one non-negligible Auger coefficient or that are not very far from a resonant LO phonon condition may effectively capture carriers.

Let us finally stress that the models discussed in this chapter refer to capture *per dot*. From the viewpoint of a single QD, as probed in spatially resolved experiments, the evaluated rates describe the loading of the dot by carriers from a reservoir outside it, under the somewhat stringent conditions discussed in section 2.1.2. Many deviations from such conditions do nevertheless often occur in practice. Their discussion is however outside the scope of this book, which aims at focusing on a few elementary processes in these systems.

Once captured into excited orbits, the carriers have to relax down to the ground dot shell. Intra-dot relaxation is discussed in the next chapter.

References

Ferreira R and Bastard G 1999 Phonon-assisted capture and intra-dot Auger relaxation in quantum dots *Appl. Phys. Lett.* **74** 2818

Magnusdottir I 2002 Modeling of phonon- and Coulomb-mediated capture processes in quantum dots *PhD thesis* Technical University of Denmark (*orbit.dtu.dk/services/downloadRegister/5254694/im_thesis_corr.pdf*)

Magnusdottir I, Bischoff S, Uskov A V and Mork J 2003 Geometry dependence of Auger carrier capture rates into cone-shaped self-assembled quantum dots *Phys. Rev. B* **67** 205326

Magnusdottir I, Uskov A V, Ferreira R, Bastard G, Mørk J and Tromborg B 2002 Influence of quasibound states on the carrier capture in quantum dots *Appl. Phys. Lett.* **81** 4318

IOP Concise Physics

Capture and Relaxation in Self-Assembled
Semiconductor Quantum Dots
The dot and its environment
Robson Ferreira and Gérald Bastard

Chapter 3

Energy relaxation of confined carriers in self-assembled quantum dots

In this chapter we discuss the energy relaxation of carriers placed in excited states of a QD. On fairly general grounds, any mechanism leading to an irreversible descent of the carrier to the ground state implies coupling of this carrier to a reservoir, capable of receiving the energy that is involved in the relaxation process. Two mechanisms will be discussed: one that implies interaction between electrons and LO phonons, and the other involving Coulomb interaction between bound carriers in the QD.

We discussed in chapter 1 the existence of a peculiar coupling between the charged carriers (electrons and holes) and LO phonons in QDs. It gives rise to the formation of mixed elementary excitations: the polaron states. We shall handle in the first part of this chapter the problem of intra-dot relaxation in the framework of electronic polarons. We shall show in a first step that the bulk-like anharmonic coupling gives rise to a finite lifetime for the polaron states. In a second step, we shall discuss how this finite lifetime leads *in fine* to the energy relaxation of the excited carrier towards the ground state in the QD.

The last part of the chapter deals with the study of energy relaxation in a QD that contains two electrons in excited orbits. We shall show that the Auger relaxation that results from Coulomb scattering can, under specific conditions, very efficiently trigger the relaxation of one carrier towards the ground S state of the QD.

In order to introduce the problems linked to energy relaxation in a QD, let us come back for a while to the magneto-absorption experiments undertaken on QD ensembles (see section 1.5). As illustrated in figure 3.1, under continuous wave (i.e. stationary) excitation, the absorption measurement implicitly supposes that there exists a relaxation efficient enough: otherwise, the absorption of one photon would saturate the transition. The relaxation allows statistical handling of the absorption signal: indeed, over a long time interval (the measurement duration) each of the resonantly excited QDs undergoes a large number of absorption/relaxation processes. It should

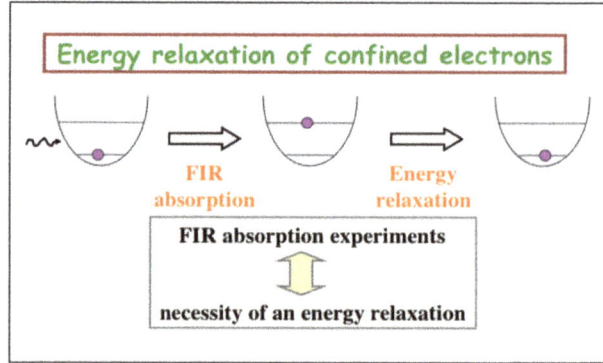

Figure 3.1. Schematic of an absorption experiment performed at low temperature in a QD. The photon energy is absorbed and the carrier is brought into an excited state. Then, it has to lose this extra energy to become available for a further absorption event.

also be stressed that the excitation is radiative while the relaxation is not (it can be shown that the spontaneous emission rate is negligibly small). Actually, the intensity of the optical excitation is often weak (characteristic strength of the radiative coupling $\ll \hbar/\tau_{rel}$, with τ_{rel} the relaxation time evaluated later in this chapter), implying a linear regime for the optical absorption versus the laser pump.

Time-resolved absorption experiments (pump-probe) conducted by Sauvage *et al* (2002) in the FIR regime have revealed the population relaxation time of the excited states of a inhomogeneous ensemble of QDs. As can be seen in figure 3.2, these times are of the order of a few tens of picoseconds and slightly (but significantly) increase with the energy of the FIR photon. Only the dots of the ensemble that have an S–P transition energy equal to that of the monochromatic excitation ($\Delta_{SP} \approx h\nu$) contribute to the absorption. Since the dots essentially have similar sizes, the measured energy variation of the relaxation time is mainly associated with dot size variation (one says that the dots present correlated geometrical and spectral distributions).

Hence, it remains to be established what the physical mechanism is that triggers this energy relaxation from the excited towards the ground dot level.

Figure 3.3 presents a partial account of the early physical interpretations of the energy relaxation in QDs. We shall not go through all the different works (quoted in the reference list at the end of this chapter Bockelmann and Bastard (1990); Bockelmann and Egeler (1992); Benisty *et al* (1991); Ferreira and Bastard (1999); Inoshita and Sakaki (1992); Li *et al* (1997); Li and Arakawa (1998); Li *et al* (1999)), but they nevertheless clearly illustrate the important efforts in the 1990s towards understanding energy relaxation in QDs. Here, we shall instead consider phonon-assisted and Coulomb-mediated processes in more detail. Let us first introduce the so-called 'phonon bottleneck' issue in QDs.

3.1 The phonon bottleneck in quantum dots

At the time of the early time-resolved *intraband* FIR experiments of Sauvage *et al* (2002) (see figure 3.2), the problem of energy relaxation in QDs had been very controversial for a long time. This was particularly due to the fact that people wanted

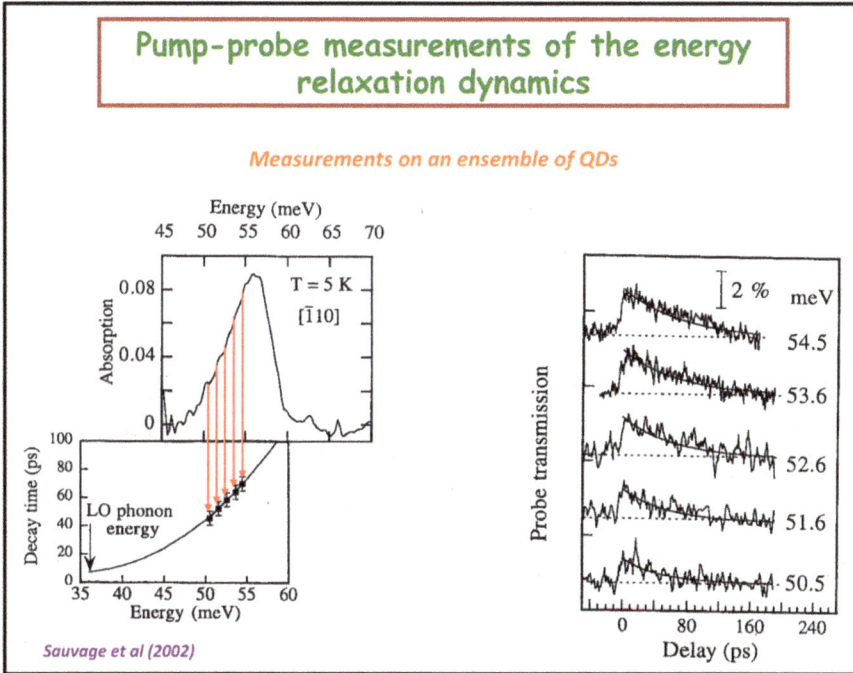

Pump-probe measurements of the energy relaxation dynamics

Measurements on an ensemble of QDs

Sauvage et al (2002)

Figure 3.2. Left upper panel: linear photon absorption associated with S–P transitions in (Ga,In)As self-assembled QDs. The FIR line is polarized parallel to one of the preferred directions of the slightly anisotropic QDs. Right panel: time decay of the probe transmission versus delay between this probe and a pump. Left lower panel: relaxation time of the population of the excited state of an ensemble of QDs for several values of the FIR photon energy. Adapted fom Sauvage *et al* (2002).

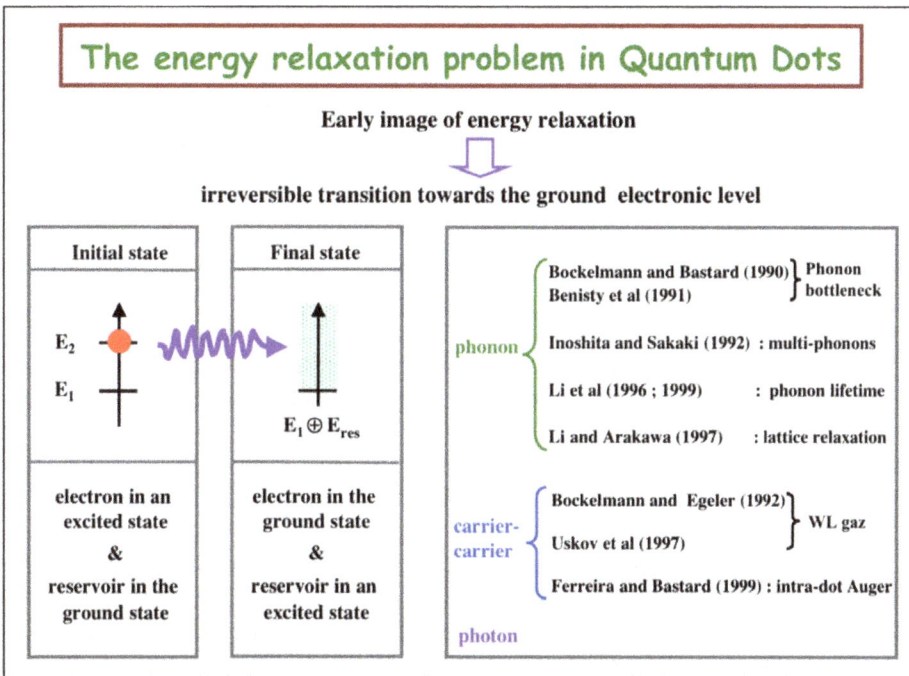

The energy relaxation problem in Quantum Dots

Early image of energy relaxation

irreversible transition towards the ground electronic level

Figure 3.3. Scheme of different mechanisms proposed to interpret the energy relaxation of carriers in QDs.

to interpret *interband* optical experiments. Several couplings were envisioned to explain the electron de-excitation from an excited state towards the ground state (see figure 3.3). FIR *intraband* experiments provided a much clearer framework to study energy relaxation in the dots. In this context, the Coulombic couplings between carriers are not so important, since doping is controlled such that the QDs are on average loaded by only one electron, and since the non-intense FIR light is unable to further load the QD by interband excitation. In addition, one may neglect the radiative P–S channel (in spite of the large electrical dipole moment of this transition). Finally, the 'usual' (in bulk and quantum wells) relaxation path, i.e. assisted by the emission of phonons, must be excluded in QDs. Let us develop this latter point below.

The first calculations of the emission rates of optical or acoustical phonons were undertaken in the weak coupling regime; that is to say, by using the Fermi golden rule. The rule's validity was firmly established in bulk materials (3D) as well as in quasi-2D electron gases, as found in multiple quantum wells. As we show below, its application to quantum dots (0D) deserves some discussion. In the weak coupling regime, the transition rate for the state $|n\rangle$ is computed as

$$\frac{1}{\tau_n} = \frac{2\pi}{\hbar} \sum_{n',\vec{Q}} \left| \langle n| \otimes \langle 0_{\vec{Q}}|H_{\text{e-ph}}|n'\rangle \otimes |1_{\vec{Q}}\rangle \right|^2 \delta\left[E_n - E_{n'} - \hbar\omega\left(\vec{Q}\right) \right], \qquad (3.1)$$

where we have considered the emission of a phonon with wavevector $\vec{Q}$ by an electron in the initial state $|n\rangle$ with no phonons; the summation runs over all the final states $|n'\rangle$ and phonon wavevectors. Acoustical and optical phonons are considered below

Let us first investigate acoustical phonon emission. The energy of one longitudinal acoustical (LA) phonon is limited: $0 < \hbar\omega_{\text{LA}}(\vec{Q}) \leqslant \Delta_{\text{LA}}$ with $\Delta_{\text{LA}} \approx 20$ meV in GaAs (see figure 1.17). Thus, if the bandwidth Δ_{LA} is smaller than $\Delta_{\text{SP}} = E_{\text{P}} - E_{\text{S}}$, then it is impossible to find any acoustical phonon wavevector that would have made the argument of the delta function vanishing. In contrast, emission becomes possible when $\Delta_{\text{LA}} > \Delta_{\text{SP}}$. Figure 3.4 shows the calculated relaxation time in a model QD, namely whose confinement is an anisotropic harmonic oscillator (we shall not provide details of the calculations here). The oscillation frequency along z is considerably larger than that in the xOy-plane ($\Omega_Z \gg \Omega_{//}$), a feature that mimics the pancake aspect of InAs QDs. The relaxation is between one state of the first excited shell (which is twofold degenerate: 1_x0_y or 0_x1_y) and the ground 0_x0_y state. Finite temperature effects have been considered; one readily establishes that when $T > 0$ and phonons are present in the initial state, the resulting de-excitation frequency $1/\tau_n$ should be multiplied by $(1+n_B(\Delta_{\text{SP}};T))$ with $n_B(\Delta_{\text{SP}}; T) = \{e^{\Delta_{\text{SP}}/k_{\text{B}}T} - 1\}^{-1}$ the Bose factor. The relaxation time is sizeable for very large dots (small in-plane quantization energy $\hbar\Omega_{//}$), and presents a drastic increase for $\hbar\Omega_{//}$ larger than a few milli-electron volts.

In self-assembled dots one has $\Delta_{\text{SP}} \sim$ a few tens of meV. Similar calculations show that the energy relaxation rate by acoustical phonon emission is either extremely improbable (if $\Delta_{\text{SP}} \leqslant \Delta_{\text{LA}}$) or becomes impossible (if $\Delta_{\text{SP}} > \Delta_{\text{LA}}$) in typical samples. In particular, in the case of the experiments in figure 1.13 one has $\Delta_{\text{SP}} \approx 50$ meV $> \Delta_{\text{LA}}$.

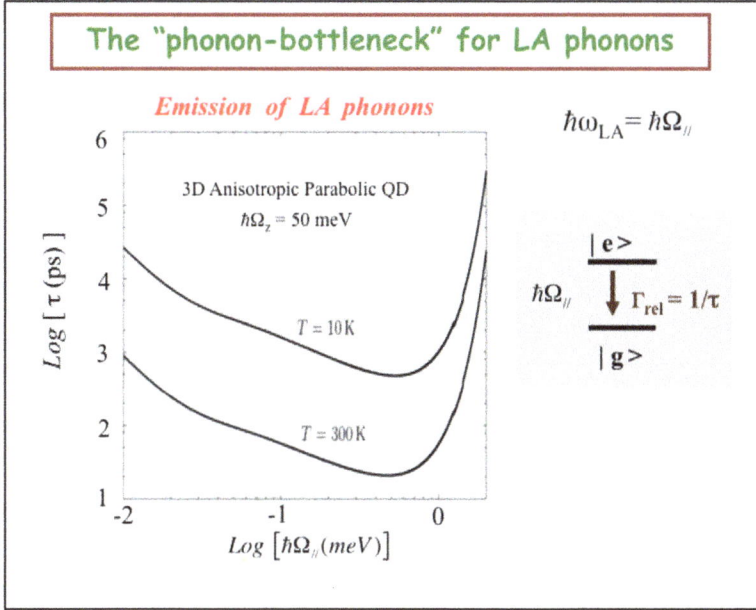

Figure 3.4. The relaxation time due to the emission of longitudinal acoustical phonons is plotted as a function of the energy separation $\hbar\Omega_{//}$ of the in-plane harmonic oscillator. From Ferreira and Bastard in Rossi and Zanardi (1995).

Let us now apply equation (3.1) to the optical phonon emission in 0D structures. Like for acoustical phonons, the energy of one LO phonon is limited: $0 < \hbar\omega_{LO}(\vec{Q}) \leqslant \Delta_{LO}$ with $\Delta_{LO} \approx 8$ meV in GaAs (see figure 1.17). Correspondingly, one has for the P–S relaxation rate,

$$\frac{1}{\tau_{SP}} = \frac{2\pi}{\hbar} \sum_{\vec{Q}} \left| \langle P | \otimes \langle 0_{\vec{Q}} | H_{e-ph} | S \rangle \otimes |1_{\vec{Q}} \rangle \right|^2 \delta \left[\Delta E_{SP} - \hbar\omega_{LO}(\vec{Q}) \right] \qquad (3.2)$$

with H_{e-ph} the Fröhlich coupling in equation (1.29). Very often we neglect this small dispersion of the LO phonons and use $\omega_{LO}(\vec{Q}) \approx \omega_{LO}(0) \equiv \omega_{LO}$. Thus, without any calculation, we see that the discretization of the low-energy spectrum of QDs leads to the impossibility of LO phonon emission using the traditional approach because there is no reason why the electronic energy difference ΔE_{SP} should match the LO phonon energy. Restoring the phonon dispersion does not help much because firstly it is very narrow and secondly the matrix elements $\langle P | H_{e-ph} | S \rangle$, coming from the Fröhlich coupling in equation (1.29), decay very rapidly with Q once $Q_x L_x$, $Q_y L_y$, $Q_z L_z$ become larger than 1 (with $L_{x,y,z}$ typical sizes of the QD), thereby very effectively shrinking the effective Q range of LO phonons that can interact with the electrons bound to the QD.

The impossibility or quasi-impossibility of LO and LA phonon emission by electrons bound to excited states of QDs has been termed the 'phonon bottleneck'; see figure 3.5. It should have a very clear signature in the interband optical spectra

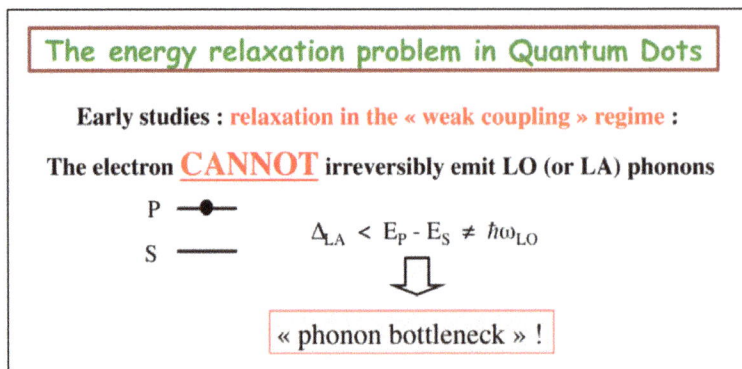

Figure 3.5. Conditions leading to the existence of a phonon bottleneck in QDs. Δ_{LA} is the width of the acoustical phonon branch.

but, despite the fact that many groups have repeated the experiments, no clear trend has been observed.

Other mechanisms have been put forward to explain why such a clear-cut phenomenon of inhibition of energy relaxation was not observed (see figure 3.3):

- The emission of two phonons, one acoustical and the other optical, relieves the monochromaticity of the LO branch. It is however effective only in a tiny energy window around the LO phonon energy.
- The relaxation mechanism associated with multi-phonon emission around a trap situated near the QD. It involves a double electronic transfer; the electron leaving the QD to the trap, relaxing within the trap and finally being recaptured. This is an extrinsic mechanism that crucially depends on the immediate surroundings of the QD.
- The mechanism we shall retain later on in this chapter is the finite lifetime of the LO phonons due to lattice anharmonicity. It leads to disintegration of the polaron states.

Before we analyze the latter point in some detail, one should mention that interband optical experiments are difficult to interpret because they necessarily involve several (at least two) carriers. These carriers can relax independently and on different time scales (because the electron and hole states bound to the QD are different), or they can relax in a correlated fashion due to Coulombic interactions (this mechanism will be discussed later in the chapter when we shall deal with the intra-QD Auger effect). Finally, a further difficulty that hampers understanding of energy relaxation is the inhomogeneous broadening of interband lines observed on ensembles of QDs that display size dispersion.

Intraband experiments offer less inhomogeneous broadening and are easier to interpret because they involve a single type of carrier. In addition, theoretical understanding of the energy levels and the electron–phonon coupling is better (the conduction bands of bulk InAs and GaAs are orbitally non-degenerate while the valence band is fourfold degenerate). Hence, intraband experiments have allowed a new framework to be created in order to understand the energy relaxation in QDs.

3.2 Polaron coupling versus phonon anharmonicity

The phonon bottleneck prediction results from an incorrect use of a weak coupling approach to describe the interaction between LO phonons and electrons bound to the QD. Actually, we know that electrons and phonons are in the strong coupling regime and that they give rise to the formation of mixed elementary excitations: the polarons (see section 1.5). This does not relieve our difficulties in understanding the energy relaxation because polarons take into account the electron–phonon coupling to any order, and the resulting polaron eigensolutions are stationary states. Hence, there is no time evolution of the system within the polaron framework, and therefore no energy relaxation.

However, when building the polaron states in section 1.5 we assumed the very existence of phonons. The latter are the elementary excitations of a crystal lattice vibrating harmonically, i.e. the restoring forces are linear in the atomic displacements and the potential energy is a quadratic function of these displacements. But this is only an approximation since there is no reason why the inter-atomic potentials should always be a quadratic function of the displacements. The difference between the true potential energy and its harmonic approximation (the anharmonicity) means that the phonons, eigenstates of the harmonic crystal, become unstable. For instance, two phonons can 'collide' and merge to form a third phonon. The reverse also applies: one phonon can 'dissociate' and give rise to two phonons. We shall be particularly interested in the last process (the only one possible for optical phonons at low temperatures). The anharmonic coupling is thus a third-order process, i.e. involving three phonons. For the processes we are interested in, its general form in bulk is

$$V_{\text{ph-ph}} = \sum_{\vec{q},\vec{Q}_1,\vec{Q}_2} \left[V_{\vec{q}\rightarrow(\vec{Q}_1,\vec{Q}_2)} \, a_{\vec{q}} a_{\vec{Q}_1}^{\pm} a_{\vec{Q}_2}^{\pm} + V_{(\vec{Q}_1,\vec{Q}_2)\rightarrow\vec{q}} \, a_{\vec{q}}^{+} a_{\vec{Q}_1} a_{\vec{Q}_2} \right] \delta_{\vec{q}-\vec{Q}_1-\vec{Q}_2} \equiv \sum_{\vec{q},\vec{Q}_1,\vec{Q}_2} V_{\vec{q},\vec{Q}_1,\vec{Q}_2},$$

$$(3.3)$$

where a^+ and a are creation and annihilation operators such that the first term in the sum represents the annihilation of one LO phonon of wavevector $\vec{q}$ with the creation of two phonons of wavevectors $\vec{Q}_1$ and $\vec{Q}_2$, and the second term describes the reverse process. The Kronecker delta accounts for conservation of total impulsion, due to bulk translational invariance. A zone-centre LO phonon can thus only emit two phonons with opposite wavevectors. We show in figure 3.6 some of the possible paths for LO phonon disintegration.

Anharmonicity gives rise to a finite lifetime for the LO phonons: it takes some 10 ps (2 ps) for a zone-centre optical phonon in GaAs to dissociate at $T = 4$ K (300 K). However, we neglected the finite phonon lifetime in equations (3.1) and (3.2) to compute the scattering time of electrons within the Fermi golden rule framework. This can be justified for the calculation of relaxation rate in bulk and quantum wells; indeed, in these materials the electronic lifetime (sub-picosecond at low temperature) is much shorter than the phonon lifetime, so that it makes sense to assume that the phonons are unperturbed by anharmonicity since they appear stable on the time scale of the electronic relaxation. In contrast, for QDs one cannot rely on a fast electronic

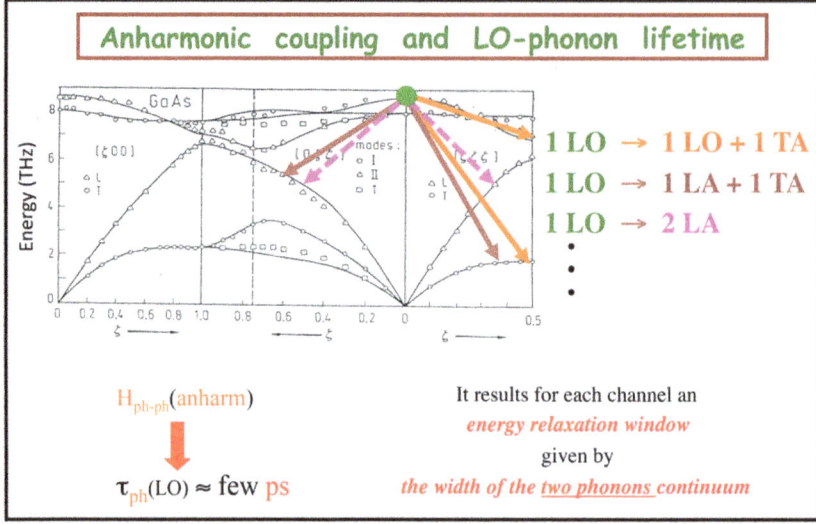

Figure 3.6. Phonon dispersion relations in bulk GaAs. The arrows exemplify possible paths for the disintegration of a zone-centre LO phonon.

relaxation to neglect the phonon lifetime since there is no relaxation at all. Thus, equation (3.2) does not apply and we need to make a better calculation to include the phonon–phonon interaction within the polaron framework. Actually, on more general grounds one should provide a model incorporating both the interaction of the electron to harmonic LO phonons by the Fröhlich coupling, and of the LO phonons to different two-phonon modes by the anharmonic coupling (this is discussed in the next sections). However, in order to roughly estimate which of the anharmonicity and Fröhlich couplings produce the largest effect on the dot states, we note that the polaron anticrossings are several milli-electron Volts wide (see, for example, figure 1.26). Thus, with a broadening of 0.3 meV or less due to anharmonicity-triggered lifetime, as shown below, one may safely take the polaronic description as the backbone of the system and study within this framework the effects of anharmonicity.

3.3 The existence of energy windows associated with anharmonic decay

Let us consider the anharmonic effects in bulk materials and concentrate initially on the path $1LO \rightarrow 1LO + 1TA$ for the sake of definiteness (see figure 3.7). A salient feature is the existence of a two-phonon continuum with a finite width. In effect, there is

$$E_{2ph}^{min}(q_{LO}, q_{TA}) = E_{2ph}(q_{BZ}, 0) = \hbar\omega_{LO} - \Delta_{LO} \approx 28 \text{ meV}$$

$$E_{2ph}^{max}(q_{LO}, q_{TA}) = E_{2ph}(0, q_{BZ}) = \hbar\omega_{LO} + \Delta_{LO} \approx 44 \text{ meV}, \quad (3.4)$$

where q_{BZ} is the Brillouin-zone edge. This is a consequence of the finite width of each one-phonon continuum (see figure 3.6): the lower (higher) two-phonon energy corresponds to one zone-edge (-centre) LO and one zone-centre (-edge) TA phonon. For another path, the energy window can be much wider: for example, the energy

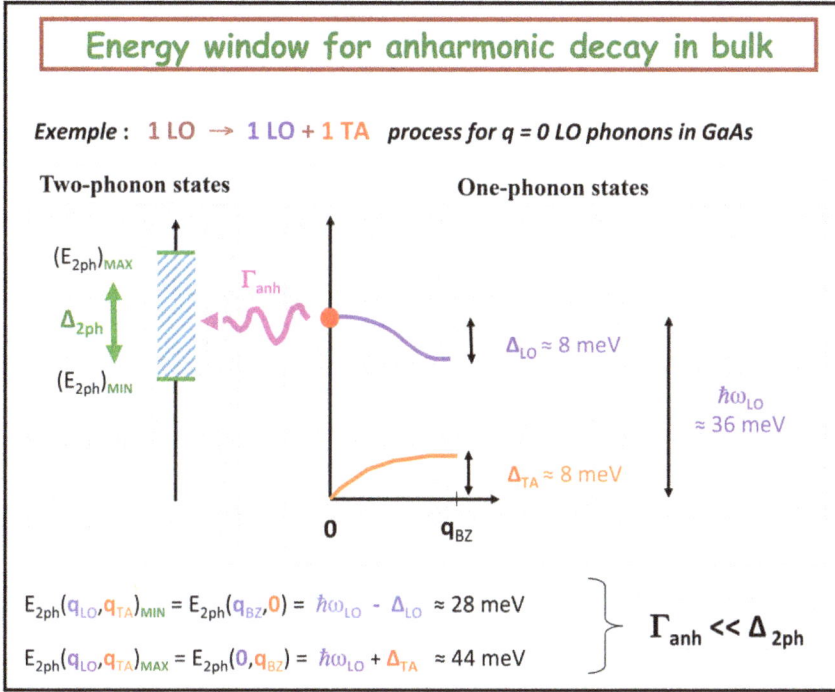

Figure 3.7. One- and two-phonon continua associated with LO and TA phonons.

window for the $1LO \rightarrow 2LA$ is [0 meV, 53 meV]. Note that these estimates are upper values, which do not take the conservation of the phonon wavevector into account.

The anharmonic couplings are weak and the Δ_{2ph} width of the two-phonon continuum remains much larger than the energy broadening $\Gamma_{anh} = \hbar/\tau_{LO}$ of the one-phonon states, where τ_{LO} is the lifetime of the $q = 0$ phonon. In addition, the initial state is, for all the possible paths, located far away in energy from the boundaries of the two-phonon continua. These conditions ensure that the Fermi golden rule is justified to compute τ_{LO} in bulk material.

In contrast, for QDs the existence of a finite energy window for the LO phonon disintegration into two other phonons may have significant consequences on the energy relaxation. To illustrate this point, figure 3.8 shows how the polaron states are positioned with respect to the two-phonon continuum associated with the $1LO \rightarrow 1LO + 1TA$ disintegration channel. Let us first consider the decoupled states $|E\rangle \otimes |N_{ph}\rangle \equiv |EN_{ph}\rangle$. The fundamental state is $|S0_{SP}\rangle$ (not shown).

The dashed lines represent the variations of the energies of the decoupled one-phonon $|S1_{SP}\rangle$ and zero-phonon $|P_{\pm}0_{SP}\rangle$ states of a dot with circular basis, as a function of its basal radius, in the radius region where these two decoupled states cross ($R \approx 130$ Å; see figure 1.26). The energies of the two polaron states $|1\pm\rangle$ and $|2\pm\rangle$ are also shown as full lines. The two-phonon continuum of states is $|\overline{S0}\rangle \otimes |2_{ph}\rangle$ (green strip), where $|\overline{S0}\rangle$ is the ground polaron state of the QD (not shown). The anharmonic potential couples the $|\pm1\rangle \otimes |0_{ph}\rangle$ and $|\pm2\rangle \otimes |0_{ph}\rangle$ excited polaron states with the $|\overline{S0}\rangle \otimes |2_{ph}\rangle$ ones. We see that for a large range of dot parameters, one of the polaron states is found outside the disintegration continuum.

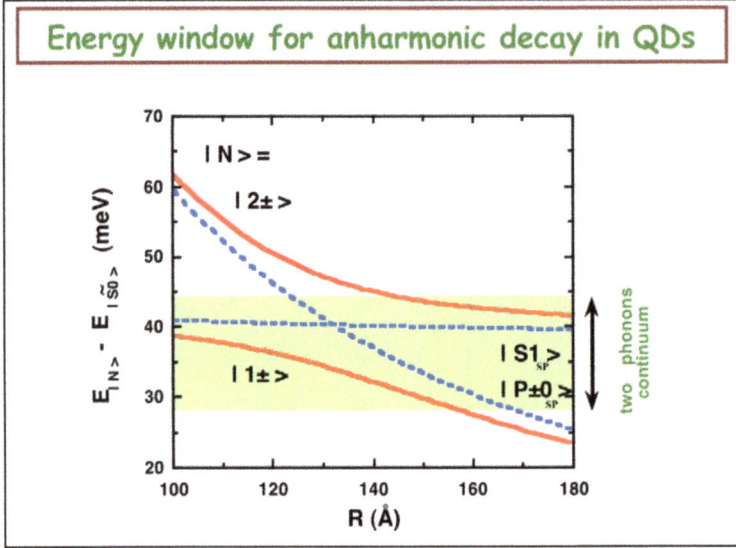

Figure 3.8. Energy window for the LO phonon disintegration path 1LO → 1LO + 1TA in bulk (green strip), compared to the decoupled states (dashed lines) and polaron energies (full lines) in a dot, as a function of its basal radius. (See also figure 1.26.)

3.4 Modelling of unstable polarons: main issues

Strictly speaking and as already mentioned, the complete description of polaron states should concomitantly account for the two phonon-related couplings (see figure 3.9): the Fröhlich coupling, which mixes the zero- and one-phonon decoupled states and leads to the formation of entangled stable states, the polarons; and the anharmonic couplings, which lead to the dissociation of the one-phonon 3D states $|1_q\rangle$ and so of their $|1_{SP}\rangle$ linear combination, and at the end to the instability of the decoupled state $|S1_{SP}\rangle$ that enters in the formation of the polaron states.

Additionally, the model description should cope with the particularities of the two-phonon states: various disintegration channels with different structures (coupling strengths, energy widths and density of states,...). Let us finally make two remarks concerning the decoupled one- and zero-phonon dot states. First, as shown in chapter 1, the electron–phonon coupling introduces a particular one-phonon mode, $|1_{SP}\rangle$. The anharmonic coupling is weak and we assume that it can be treated in perturbation: $\hbar/\tau_{SP} \ll \Delta_{2ph}$, where τ_{SP} is the lifetime of $|1_{SP}\rangle$ related to the channel of width Δ_{2ph}. Thus, like for the bulk case, the fact that the channel is of finite width is of no relevance in the evaluation of the lifetime of the *decoupled* one-phonon state $|S1_{SP}\rangle$. Additionally, note that the $|1_{SP}\rangle$ state is made of bulk phonons of $q \neq 0$ wavevectors, which have a different lifetime from the value τ_{LO} for the centre zone ($q = 0$) mode, so that in general $\tau_{SP} \neq \tau_{LO}$ (we shall come back to this distinction later).

Second, the zero-phonon decoupled states are not directly affected by the anharmonicity, but only indirectly via their coupling to the one-phonon states. This situation reminds us of the situation in atomic physics in vacuum, where an excited stable (or

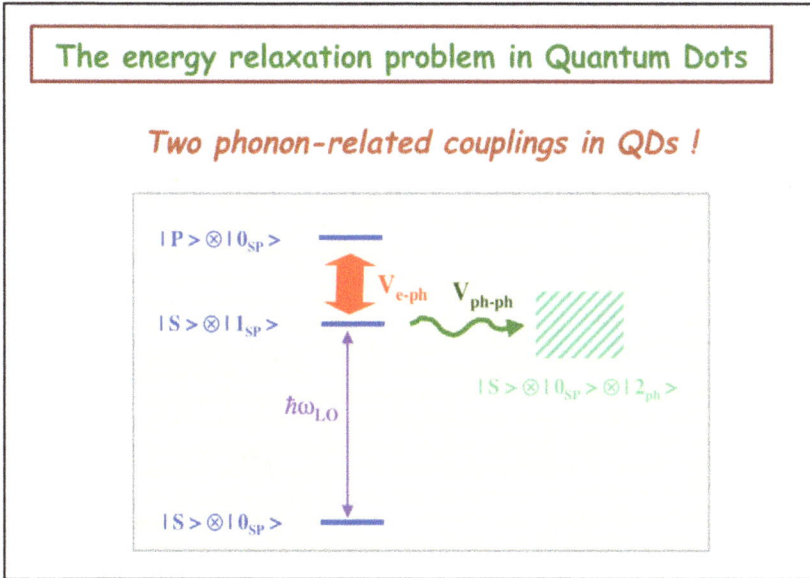

Figure 3.9. Scheme showing the problems associated with energy relaxation in QDs in the presence of Fröhlich (red arrow) and anharmonic (green arrow) couplings.

metastable) state acquires a finite lifetime by coupling ('contamination') with a radiative state, of either higher or lower energy, in the presence of, e.g., a radio-frequency excitation. The latter role in our case is played by the Fröhlich coupling. However, for an atom in vacuum the metastable state is always immersed in the disintegration continuum of states (since photons may in principle have any energy), whereas in our case $|P_{\pm}0_{SP}\rangle$ may be found outside the two-phonon continuum.

To date, two models have been proposed to handle the phonon-assisted relaxation in a QD. They both use a two-step account of the electron–phonon (V_{e-ph}) and phonon–phonon (V_{ph-ph}) couplings but differ in the *order* in which these two are considered (see figure 3.10): in the 'semiclassical' model of Li *et al* (1999) one considers first the broadening of the one-phonon dot state due to V_{ph-ph} and then turns on its Fröhlich coupling to the stable zero-phonon states; and in the 'polaron' model of Verzelen *et al* (2000) one first diagonalizes V_{e-ph} using stable zero- and one-phonon dot states, and then considers the effect of the anharmonic coupling on the resulting stationary polaron states.

The two models should in principle predict the same relaxation dynamics. Their predictions are however often in great disagreement, as we will show below.

3.5 General approach to the unstable polaron states

In the following we develop a general approach to the relaxation problem, which will also allow us to present the two previous models. The first (obvious) point to realize when constructing a model for the anharmonic-driven mechanism is that we should incorporate the two-phonon states $|2_{ph}\rangle$ in the description. The working basis involving the ground (1S) and one excited (1P) orbitals becomes $|P, 0_{LO}\rangle \otimes |2I\rangle$, $|S, 1_{\vec{q}}\rangle \otimes |2I\rangle$ and $|S, 0_{LO}\rangle \otimes |2F\rangle$, where $|0_{LO}\rangle$ and $|1_{\vec{q}}\rangle$ represent the vacuum and

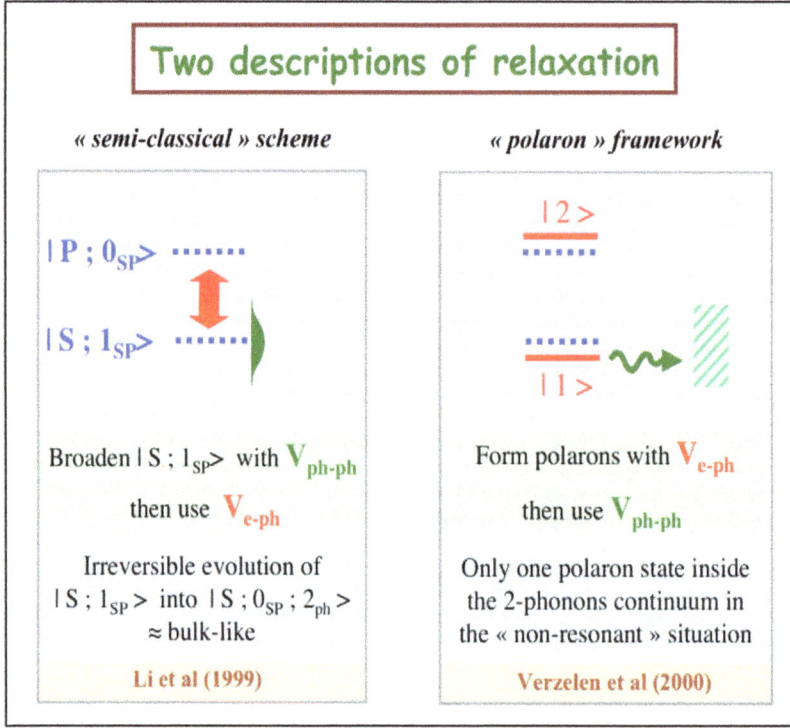

Figure 3.10. Comparative scheme for the two models used to tackle the energy relaxation problem in QDs: the 'semiclassical' (left) and 'polaron' (right) models.

one-$\vec{q}$ LO phonon states, and $|2I\rangle$ and $|2F\rangle$ are the initial and final states of the two-phonon reservoir that differ by two arbitrary low-energy phonons (those of wave-vectors $\vec{Q}_1$ and $\vec{Q}_2$ in equation (3.3)). In this basis there is

$$|\Psi(t)\rangle = \left[A(t)|P; 0_{LO}\rangle + \sum_{\vec{q}} B_{\vec{q}}(t)|S; 1_{\vec{q}}\rangle \right] \otimes |2I\rangle$$

$$+ |S; 0_{LO}\rangle \otimes \sum_{2F} C_{2F}(t)|2F\rangle \qquad (3.5a)$$

$$\Rightarrow \begin{cases} i\hbar \dfrac{\partial A}{\partial t} = \left(E_P + E_{2I} \right) A + \sum_{\vec{q}} V_{\vec{q}} B_{\vec{q}} \\[2mm] i\hbar \dfrac{\partial B_{\vec{q}}}{\partial t} = \left(E_S + \hbar\omega_{LO} + E_{2I} \right) B_{\vec{q}} + V_{\vec{q}}^* A + \sum_{2F} V_{2F,0_{\vec{q}}}^{2I,1_{\vec{q}}} C_{2F} \qquad (3.5b) \\[2mm] i\hbar \dfrac{\partial C_{2F}}{\partial t} = \left(E_S + E_{2F} - i\gamma_{2F} \right) C_{2F} + V_{2I,1_{\vec{q}}}^{2F,0_{\vec{q}}} B_{\vec{q}}, \end{cases}$$

where the LO phonons are taken to be dispersionless, $V_{2F,0_{LO}}^{2I,1_{\vec{q}}} = \langle S; 1_{\vec{q}};$ $2I|V_{\vec{q},\vec{Q}_1,\vec{Q}_2}|S; 0_{LO}; 2F\rangle$ is the matrix element of the anharmonic potential, $E_{2I} = E(|2I\rangle)$, $E_{2F} = E(|2F\rangle)$) and a phenomenological broadening $-i\gamma_{2F}$ has been added to the final two-phonon states. Note that the sum over $2F$ actually represents a sum over all $\vec{Q}_1$ and $\vec{Q}_2$. This system generalizes the one in equation (1.35) for stable polarons. Let us then seek solutions in the form $\exp\{-i(E+E_{2I})t/\hbar\}$. We are led to solve the system

$$
\begin{cases}
(E_P - E)A + \sum_{\vec{q}} V_{\vec{q}} B_{\vec{q}} = 0 \\[2mm]
\left(E_S + \hbar\omega_{LO} - E\right)B_{\vec{q}} + V_{\vec{q}}^* A + \sum_{2F} V_{2F,0_{LO}}^{2I,1_{\vec{q}}} C_{2F} = 0 \\[2mm]
\left(E_S + \Delta E_{2F} - i\gamma_{2F} - E\right)C_{2F} + V_{2I,1_{\vec{q}}}^{2F,0_{LO}} B_{\vec{q}} = 0,
\end{cases}
\tag{3.6}
$$

where $\Delta E_{2F} = E_{2F} - E_{2I}$ is the energy variation of the two-phonon reservoir. The solutions of this system fulfil

$$
E_P - E + \sum_{\vec{q}} \frac{|V_{\vec{q}}|^2}{E - \left[E_S + \hbar\omega_{LO} + F_{\vec{q}}(E)\right]} = 0
$$

$$
F_{\vec{q}}(E) = \sum_{2F} \frac{\left|V_{2F,0_{LO}}^{2I,1_{\vec{q}}}\right|^2}{E - \left(E_S + \Delta E_{2F} - i\gamma_{2F}\right)}.
\tag{3.7}
$$

Owing to the weakness of V_{ph-ph}, one makes the approximation

$$
F(E) \to \text{Im}\left[\lim_{\gamma_{2F} \to 0} F(E)\right] = -i\frac{\hbar}{2\tau_{QD}(E; \vec{q})} \equiv -i\Gamma_{QD}(E; \vec{q})
$$

$$
\frac{1}{\tau_{QD}(E; \vec{q})} = \frac{2\pi}{\hbar} \sum_{2F} \left|V_{2F,0_{LO}}^{2I,1_{\vec{q}}}\right|^2 \delta\left[E - E_S - \Delta E_{2F}\right].
\tag{3.8}
$$

The rate $\Gamma_{SP}(E; \vec{q})$ is the key quantity in the study of energy relaxation in QDs. It depends on many parameters: the anharmonic channel (via $|2I\rangle$ and $|2F\rangle$ states in the matrix elements and ΔE_{2F}); the wavevector $\vec{q}$ of the LO phonon; the QD parameters (via the $|S\rangle$ wavefunction in the matrix elements and E_S); and the total energy E.

How such dependences are treated forms the origin of the differences between the two energy relaxation models, as discussed in the following.

3.5.1 Stationary polarons

In the absence of anharmonic coupling ($F(E) = 0$), one obtains the energies $\varepsilon_\pm$ of the two stationary polaron states, as discussed in chapter 1 (see, for example, equation (1.38)).

3.5.2 Semiclassical relaxation model

In the so-called semiclassical model one neglects the energy and wavevector dependences of the rate $\Gamma_{QD} = \hbar/(2\tau_{QD})$. This can be realized by, e.g., evaluating equation (3.8) at the energy $E = E_S + \hbar\omega_{LO}$ of the $|S; 1_{SP}\rangle$ decoupled state, and either considering some average over the wavevectors for $1/\tau_{QD}(E_S + \hbar\omega_{LO}; \vec{q})$ or evaluating it at the zone centre $q = 0$, so

$$\Gamma_{SC} = \frac{\hbar}{2\tau_{SC}}$$

$$\frac{1}{\tau_{SC}} \equiv \frac{1}{\tau_{QD}(E = E_S + \hbar\omega_{LO}; \vec{q} = 0)} = \frac{2\pi}{\hbar} \sum_{2F} \left| V_{2F,0_{LO}}^{2I,1_{\vec{q}=0}} \right|^2 \delta[\hbar\omega_{LO} - \Delta E_{2F}]. \quad (3.9)$$

Actually, and as will become clearer below, this approximation amounts to assuming a constant finite lifetime of the $|1_{SP}\rangle$ phonon mode with which the polaron states are formed. Indeed, in this case the problem in equation (3.7) becomes equivalent to proposing the semiclassical eigenproblem

$$\left| \Psi_{S-C}(t) \right\rangle = A(t)|P; 0_{SP}\rangle + B(t)|S; 1_{SP}\rangle$$

$$\Rightarrow i\hbar \frac{\partial}{\partial t}\begin{pmatrix} A \\ B \end{pmatrix} = \begin{pmatrix} E_P & V_{eff} \\ V_{eff}^* & E_S + \hbar\omega_{LO} - i\Gamma_{SP} \end{pmatrix}\begin{pmatrix} A \\ B \end{pmatrix}, \quad (3.10)$$

where the one-phonon decoupled state is considered unstable from the beginning, and subsequently diagonalizing the Fröhlich coupling, to obtain the energies $E_\pm$

$$\det\begin{pmatrix} E_P - E & V_{eff} \\ V_{eff}^* & E_S + \hbar\omega_{LO} - i\Gamma_{SP} - E \end{pmatrix} = 0$$

$$\Rightarrow E_\pm = E_S + \hbar\omega_{LO} + \frac{\Delta E - i\Gamma_{SP}}{2} \pm \sqrt{\left(\frac{\Delta E + i\Gamma_{SP}}{2}\right)^2 + V_{eff}^2} \quad (3.11)$$

with $\Delta E = E_P - (E_S + \hbar\omega_{LO})$. These energies also result from equation (3.7) using the definition of V_{eff} in equation (1.38).

Compare then $E_\pm$ with $\varepsilon_\pm$ in equation (1.38) for the stable polarons. Note that a one-half factor appears in the broadening $\Gamma_{SP} = \hbar/(2\tau_{SP})$ because the eigenproblem concerns the probability amplitude $B(t)$ of the one-phonon state rather than its occupancy. Indeed, in the absence of Fröhlich coupling there is $B(t) = B(0)\exp\{-i[\hbar\omega_{LO} + E_S - i\Gamma_{SP}]t/\hbar\}$ so that the one-phonon occupancy is $|B(t)|^2 = |B(0)|^2\exp\{-t/\tau_{SP}\}$, in agreement with the interpretation of τ_{SP}.

In the following let us focus on the predictions of this model as regards to the relaxation issue in QDs. The two solutions $E_\pm$ have an imaginary part when $\Gamma_{SP} \neq 0$. To the lowest order in Γ_{SP} one has

$$E_\pm(\Gamma_{SP}) \approx E_\pm(\Gamma_{SP} = 0) + \Gamma_{SP}\left(\frac{\partial E_\pm(\Gamma_{SP})}{\partial \Gamma_{SP}}\right)\Bigg|_{\Gamma_{SP}=0} = \varepsilon_\pm - i\Gamma_{SP}|\alpha_\pm|^2$$

$$\begin{cases} |\alpha_+|^2 = \sin^2(\theta) = \frac{1}{2}\left\{1 - \frac{\Delta E}{\sqrt{\Delta E^2 + 4V_{eff}^2}}\right\}. \\ |\alpha_-|^2 = \cos^2(\theta) \end{cases} \quad (3.12)$$

Thus, as expected, the polaron states become unstable: if one excites at $t = 0$ either the upper $|\psi(0)\rangle = |\psi_+(0)\rangle = |2\rangle$ or the lower $|\psi(0)\rangle = |\psi_-(0)\rangle = |1\rangle$ state, then its survival probability is

$$P_\pm(t) \equiv \left| \left\langle \psi_\pm(0) \big| \psi_\pm(t) \right\rangle \right|^2 = \exp(-t/\tau_\pm)$$
$$\frac{1}{\tau_\pm} = |\alpha_\pm|^2 \frac{1}{\tau_{SP}} \qquad (3.13)$$

It is worth pointing out two important aspects of this result. First, for both polaron states the decay rate is equal to the anharmonic decay rate of $|1_{SP}\rangle$, multiplied by the weight (squared value) of the $|S; 1_{SP}\rangle$ component on the initial polaron wave-function: $\sin^2(\theta)$ for $|2\rangle$ and $\cos^2(\theta)$ for $|1\rangle$ (see equation (1.43)). Second, the lifetime of the polaron state with energy far above or far below the one-phonon resonance increases continuously (quadratic law) with detuning and, in particular, remains finite even when $|\psi(0)\rangle$ is found outside the two-phonon disintegration continuum at the origin of Γ_{SP}. Indeed, at resonance the two polarons exhibit the same decay ($\tau_\pm = 2\tau_{SP}$ for $\Delta E = 0$), whereas at large detuning one has approximately

$$\frac{1}{\tau_{|Pol\rangle}} \xrightarrow[\left|\frac{\Delta E}{V_{eff}}\right| \gg 1]{} \begin{cases} \dfrac{1}{\tau_{SP}} \dfrac{V_{eff}^2}{\Delta E^2} & \text{for } |Pol\rangle \approx |P; 0_{SP}\rangle \\[2mm] \dfrac{1}{\tau_{SP}} & \text{for } |Pol\rangle \approx |S; 1_{SP}\rangle \end{cases} \qquad (3.14)$$

We shall come back later to such ΔE dependences.

It is also instructive to study the semiclassical dynamics when the initial state is the zero-phonon decoupled one, $|\psi(t = 0)\rangle = |P; 0_{SP}\rangle$, as has been discussed previously for stationary polarons (section 1.5.10). Here, the probability of finding the system in this zero-phonon state (survival probability) is

$$P_0(t) \equiv |\langle \psi(0)|\psi_{S-C}(t)\rangle|^2 = |A(t)|^2$$
$$A(t) = A_+ \exp(-iE_+t/\hbar) + (1 - A_+)\exp(-iE_-t/\hbar), \qquad (3.15)$$

where A_+ is a constant. The exponentials introduce both oscillations and decay (since $\text{Im}[E_\pm] < 0$). At long times the exponential with smaller $|\text{Im}[E_\pm]|$ dominates, and one has

$$P_0(t \gg \tau_P) \approx F_{osc}(t)e^{-t/\tau_P}$$
$$\frac{1}{\tau_P} = |\alpha|^2 \frac{1}{\tau_{SP}}$$
$$|\alpha|^2 = \frac{1}{2}\left\{ 1 - \frac{|\Delta E|}{\sqrt{\Delta E^2 + 4V_{eff}^2}} \right\}, \qquad (3.16)$$

where $F_{osc}(t)$ is a periodic function of time that can be readily deduced from the previous expressions. However, it is more interesting to focus on the decay rate at long times, which is given by the value $1/\tau_{SP}$ multiplied by the weight (squared value) of the $|S; 1_{SP}\rangle$ component of the polaron state nearer to $|P; 0_{SP}\rangle$, and so decreases continuously from 1/2 at $\Delta E = 0$ to zero at large detunings. Based on this dynamical model, Li *et al* (1999) associated the damping $1/\tau_P$ to the relaxation rate for the $|P; 0_{SP}\rangle$ population. This association leads to two (complementary) interpretations for the relaxation process in the dot:

- In the time domain, the electron initially in the $|P; 0_{SP}\rangle$ state periodically visits the $|S; 1_{SP}\rangle$ state in the presence of Fröhlich coupling. When $|1_{SP}\rangle$ is unstable, at each visit the electron has a non-vanishing probability of transferring its energy to the two-phonon reservoir: this is the so-called 'ping-pong' or 'semiclassical' picture for irreversible energy relaxation.
- In the frequency domain, the energies of the polaron states acquire an imaginary component that is equal to the lifetime of $|1_{SP}\rangle$ divided by the weight of its one-phonon component.

Before concluding, it is also worth pointing out that in the large detuning region where $\Delta E \gg \Gamma_{SP}$, $|V_{eff}|$ and the polaron admixing is weak, one has approximately

$$\frac{1}{\tau_P} \approx \frac{2\pi}{\hbar} |V_{eff}|^2 \frac{\Gamma_{SP}/\pi}{\Delta E^2 + \Gamma_{SP}^2}. \tag{3.17}$$

This expression indicates that the relaxation can be interpreted as Fröhlich-induced disintegration of the initial zero-phonon state into the 'continuum' issued from the broadened one-phonon state (see figure 3.11).

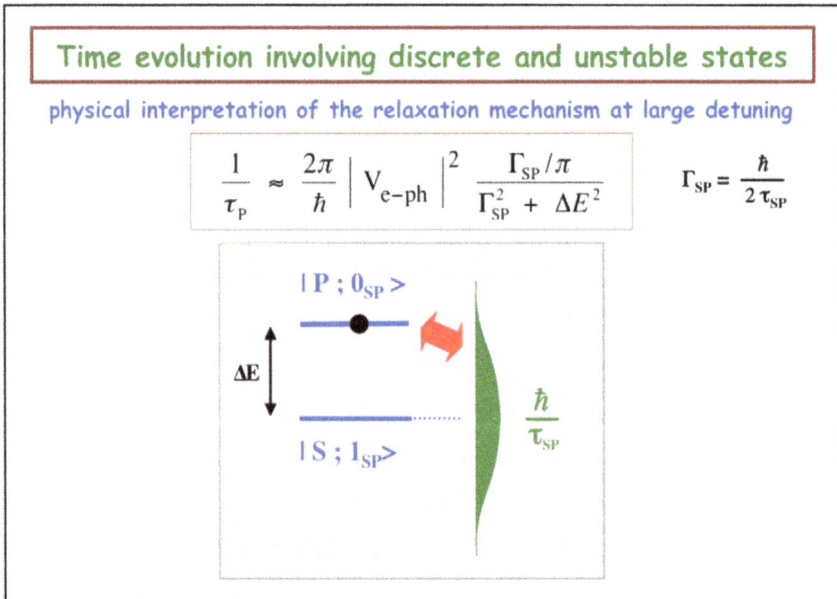

Figure 3.11. Energy relaxation at large detuning in the semiclassical model: a Fröhlich-induced dissociation into the continuum of the 1S-related broadened state.

Note finally that in the semiclassical model the constant Γ_{SP} differs from the one in the bulk material because of the QD dependence in the definition of $|1_{SP}\rangle$. One has thus in general $\Gamma_{SP} \neq \Gamma_{LO}$. As we shall see in section 3.5.4, its value has been extracted from a comparison of the above predictions with earlier experiments reported in figure 3.2.

3.5.3 Polaron relaxation model

In the polaron model one seeks a perturbative solution of equations (3.7) and (3.8) up to second order in anharmonic coupling (linear order in $F(E)$),

$$E \approx \varepsilon - i\gamma$$

$$\Rightarrow \begin{cases} \left(E_P - \varepsilon\right)\left[E_S + \hbar\omega_{LO} - \varepsilon\right] \approx V_{\text{eff}}^2 \\ \gamma + \sum_q |V_{\vec{q}}|^2 \dfrac{\gamma - \Gamma_{QD}(\varepsilon; \vec{q})}{\left[\varepsilon - E_S + \hbar\omega_{LO}\right]} \approx 0. \end{cases} \qquad (3.18)$$

It has been assumed that γ is linear in F and that $|\varepsilon - E_S + \hbar\omega_{LO}| \gg |\gamma - \Gamma_{QD}(\varepsilon; \vec{q})|$, an assumption that can be checked *a posteriori*. The first equation leads to the stable polaron energies $\varepsilon_\pm$ already discussed. The second one gives

$$\gamma_\pm = |\alpha_\pm|^2 \frac{1}{V_{\text{eff}}^2} \sum_{\vec{q}} |V_{\vec{q}}|^2 \, \Gamma_{QD}(\varepsilon_\pm; \vec{q}) \equiv |\alpha_\pm|^2 \langle \Gamma_{QD}(\varepsilon_\pm) \rangle \equiv |\alpha_\pm|^2 \, \Gamma_{SP}(\varepsilon_\pm)$$

$$|\alpha_\pm|^2 = \frac{1}{2}\left\{1 \mp \frac{\Delta E}{\sqrt{\Delta E^2 + 4V_{\text{eff}}^2}}\right\}. \qquad (3.19)$$

When $\Gamma_{QD}(\varepsilon_\pm; \vec{q})$ is approximated by a constant value Γ_{SP} (as $\Gamma_{QD}(E_S + \hbar\omega_{LO}; q = 0)$ or $\langle\Gamma_{QD}(E_S + \hbar\omega_{LO}; \vec{q})\rangle$) we recover the results of the semiclassical model in equation (3.12), as expected.

The average $\langle\Gamma_{SP}(\varepsilon_\pm)\rangle$ is the anharmonic lifetime of the one-phonon decoupled state $|S, 1_{SP}\rangle$ of the QD, *as thought* $|1_{SP}\rangle$ had the energy $\varepsilon_\pm - E_S$ and *not* its actual value $\hbar\omega_{LO}$. This result amounts to considering that the energy relaxation in the dot results from anharmonic coupling treated to the lowest order in perturbation on the stable polaron ones. Indeed, one easily applies the Fermi golden rule to obtain the disintegration rates for the stationary states $|\psi_+\rangle = |2\rangle$ and $|\psi_-\rangle = |1\rangle$ given in equation (1.43),

$$\left|\psi_\pm\right\rangle = \alpha_\pm |S; 1_{SP}\rangle + \beta_\pm |P; 0_{LO}\rangle$$

$$\Rightarrow \Gamma_\pm = \frac{\hbar}{2\tau_\pm} \equiv \pi \sum_{2F} \left|\langle\psi_\pm; 2I | V_{\text{ph-ph}} |S; 0_{LO}; 2F\rangle\right|^2 \delta\left[\varepsilon_\pm + E_{2I} - \left(E_S + E_{2F}\right)\right]$$

$$\equiv |\alpha_\pm|^2 \langle\Gamma_{SP}(\varepsilon_\pm)\rangle, \qquad (3.20)$$

where we used the fact that $V_{\vec{q},\vec{Q}_1,\vec{Q}_2}$ is proportional to $\delta_{\vec{q} - \vec{Q}_1 - \vec{Q}_2}$.

3.5.4 Comparison of polaron and semiclassical predictions

Figure 3.12 presents the results of calculations of Verzelen *et al* (2000) for the disintegration mean time (lower panel) of the two polaron states (full lines in the upper panel) as functions of the dot radius, for the 1LO → 1LO + 1TA channel. The shaded zone in the upper panel gives the energy window for relaxation (figure 3.8). We should note that in these calculations the strength $V_{\vec{q}-(\vec{Q}_1,\vec{Q}_2)}$ of the anharmonic coupling has been taken constant with a value arbitrarily adjusted such that for the weakly coupled polaron $|\psi\rangle \approx |S; 1_{SP}\rangle$ there is $\langle 1/\tau(E_S + \hbar\omega_{LO})\rangle = 0.25$ ps^{-1} ($= 1/\tau_{LO}$ in the figure) Let us make three points:

- The main difference with the semiclassical model is that the disintegration rate is evaluated at the energy of the polaron state, instead of the constant one $\hbar\omega_{LO}$ (compare equations (3.9) and (3.20)). Thus, the detuning dependence of the decay rate is no longer governed by the sole one-phonon weight $|\alpha(\Delta E)|^2$, but depends also upon the variation of $1/\tau_{SP}(\varepsilon_\pm)$, which incorporates the specificities of the particular disintegration channel under study.
- *Large detuning*. At large positive or negative detuning the lifetime of the polaron with energy near $E_S + \hbar\omega_{LO}$ is roughly $\tau(E_S + \hbar\omega_{LO}) \approx 4$ ps in both models. For large $|\Delta E|$ the other polaron state may be found outside the disintegration continuum: in this case the Fermi golden rule does not apply and this polaron is stable (regarding the 1LO → 1LO + 1TA channel and to this order of perturbation), in sharp contrast with the predictions of the semiclassical model.

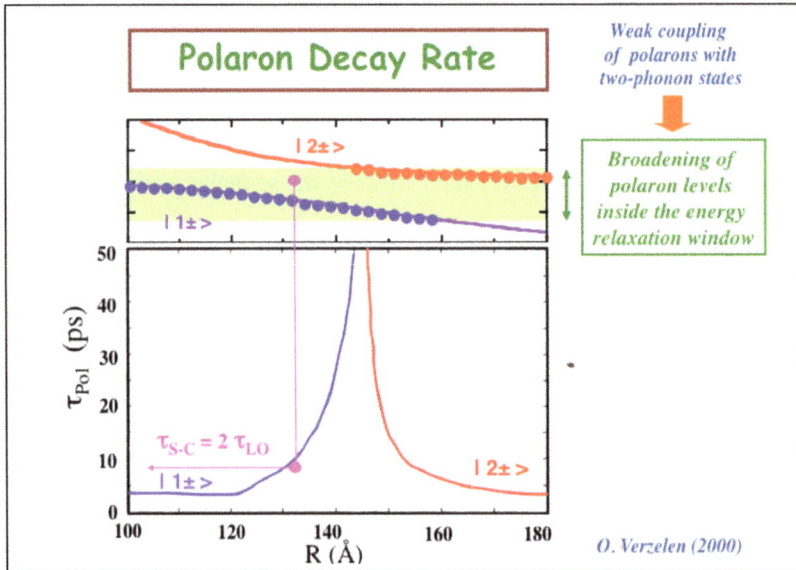

Figure 3.12. Calculated disintegration time (lower panel) of the two polaron states (full lines in the upper panel) as functions of the dot radius, for the 1LO→1LO + 1TA channel. The shaded zone in the upper panel gives the energy window for relaxation. The lines with circles (upper panel) indicate states with a finite lifetime (given in the lower panel). $T = 4$ K. See also figures 1.26 and 3.8. Adapted from Verzelen *et al* (2000).

- *Near resonance.* Around $R = 130$ Å in the figure, the predictions of both models agree for the low-energy polaron, which is well inside the two-phonon continuum: $\tau_{SP}(\varepsilon_-) \approx \tau_{SC} \approx 2\,\tau_{SP}(E_S + \hbar\omega_{LO})$. However, Fröhlich coupling pushes the upper state outside this continuum, rendering it stationary.

In conclusion, the predictions of the two models agree reasonably well for polaron states deeply inside the disintegration continuum, whereby the energy dependence of the polaron relaxation rate is weak. However, they radically differ whenever the details of the relaxation continuum become important, in particular at large detuning such that a polaron state is pushed outside this continuum.

3.5.5 Experimental evidence for many-channel relaxation

It is worth mentioning that an adjustable version of the semiclassical model was initially used with success to interpret the early measurements of Sauvage *et al* (2002) (see figure 3.2). Indeed, it showed (see the left-lower panel of figure 3.13) that the measured increase of the decay time with energy detuning nicely followed a semiclassical-like dependence, where both τ_{SP} and $|V_{e-ph}|$ were taken as free fitting parameters. A few years later, Zibik *et al* (2004) measured the decay time in a larger

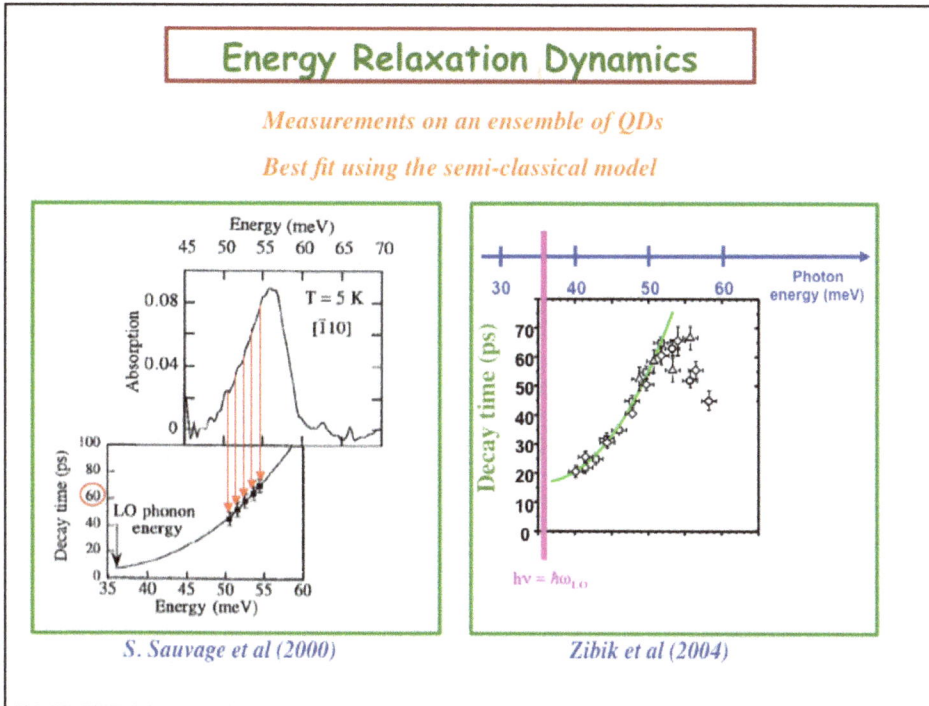

Figure 3.13. Left upper panel: S–P absorption spectrum in (Ga,In)As dots. Left lower and right panels: relaxation times of the population of the excited state of QD ensembles, for several values of the FIR photon energy: in both cases, the roughly quadratic curves are fits based on the semiclassical model. Left: adapted fom Sauvage *et al* (2002). Right: adapted from Zibik *et al* (2004).

interval of energy detuning. They pointed out that the fitting procedure worked equally well for their data in the same energy interval as Sauvage *et al* (2002); namely, $0 < \Delta E \leqslant 53$ meV. However, their results for larger detuning were in both quantitative and qualitative disagreement with a semiclassical law, even with adjustable parameters: indeed, the measured decay time *decreased* with increasing ΔE above around 53 meV (right panel in figure 3.13).

As we discuss in the following, these high detuning results clearly demonstrate the importance of an accurate description of the multiple anharmonic couplings and two-phonon reservoirs.

3.6 Many-channel relaxation of quantum dot polarons

We show in this section that the existence of various two-phonon channels (as pictured in figure 3.6) is of paramount importance for the interpretation of relaxation in QDs. The existence of many relaxation channels for polarons is inherent to the existence of many acoustical and optical bulk modes. We present schematically in figure 3.14 some of the two-phonon reservoirs for the disintegration of the LO $(q = 0)$ bulk mode. Two of them have special names: 1LO → 1LO + 1TA (Vallée-Bogani) and 1LO → 1LA + 1LA (Klemens). In bulk, the measured relaxation rate involves all possible paths, and reads

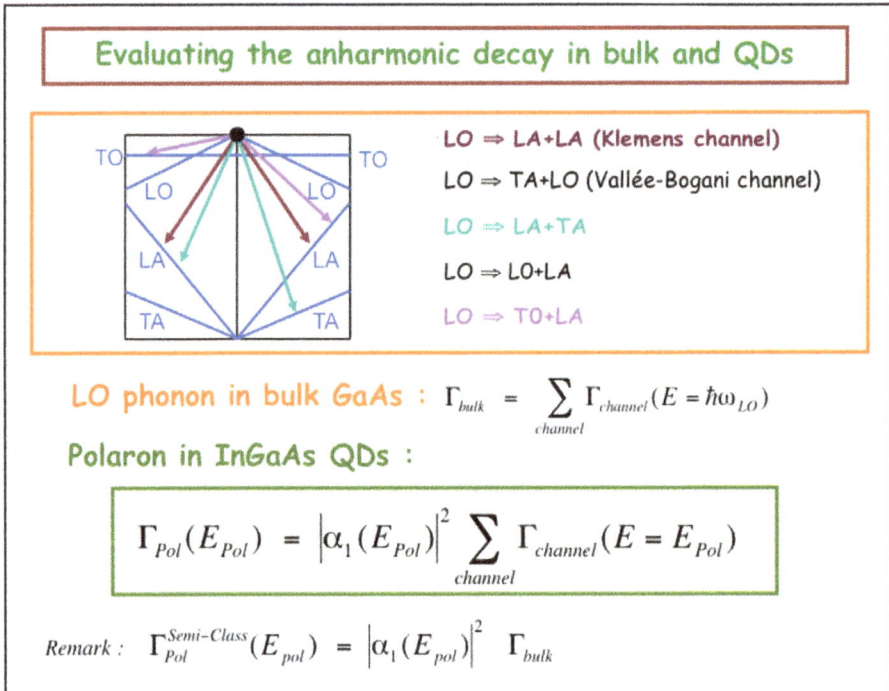

Figure 3.14. Scheme of the many channels for anharmonic disintegration of the centre-zone LO mode in bulk. The phonon dispersions have been approximated by flat or straight lines. The formulas represent the total bulk and polaron decay rates. From Grange *et al* (2007).

$$\Gamma_{\text{bulk}} = \sum_{\text{channel}} \Gamma_{\text{channel},1_{\text{LO}}}(E = \hbar\omega_{\text{LO}}). \tag{3.21a}$$

In the same way, in QDs we have

$$\Gamma_{\text{Pol}}(E_{\text{Pol}}) = \left|\alpha_1(E_{\text{Pol}})\right|^2 \sum_{\text{channel}} \Gamma_{\text{channel},1_{\text{SP}}}(E = E_{\text{Pol}}), \tag{3.21b}$$

while the semiclassical approximation reads

$$\Gamma_{\text{S–C}}(E_{\text{pol}}) = \left|\alpha_1(E_{\text{pol}})\right|^2 \sum_{\text{channel}} \Gamma_{\text{channel},1_{\text{SP}}}(E = \hbar\omega_{\text{LO}}). \tag{3.21c}$$

Each of these different paths produce new energy windows, as shown in figure 3.15. Note that bulk disintegration already involves a few different channels (vertical bar). For dot polarons, the number and efficiency of each channel will depend upon the polaron energy, as we describe in the following.

3.6.1 Relaxation of high-energy polarons

The Klemens process is of particular importance in QDs. Indeed, one clearly sees in figure 3.15 that an observed discontinuity of the decay rate around 53 meV appears

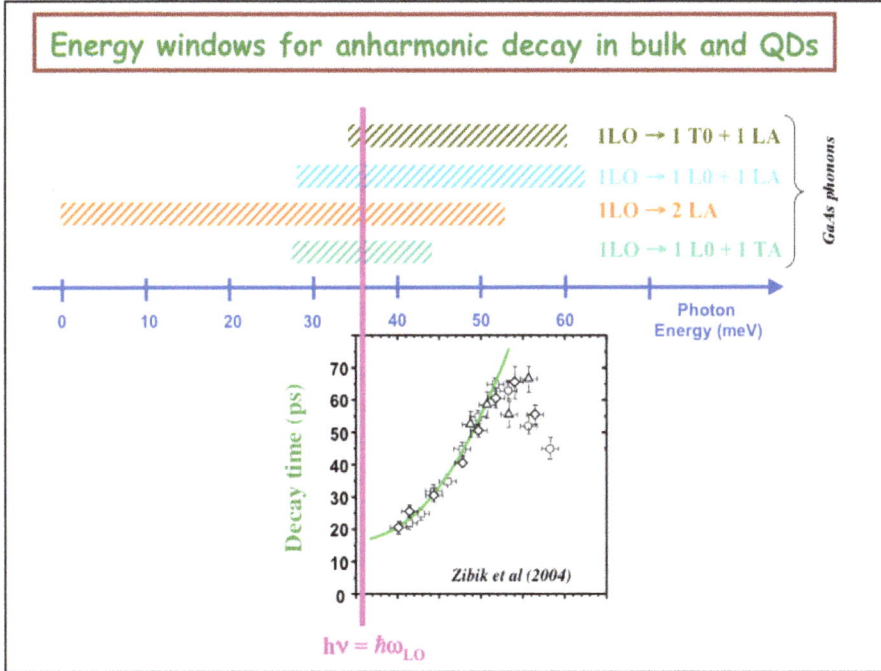

Figure 3.15. Scheme of different channel windows for anharmonic disintegration of either the centre-zone LO mode in bulk (vertical bar) or QD polaron. Lower panel: measured time decay for dot polarons. Adapted from Zibik *et al* (2004).

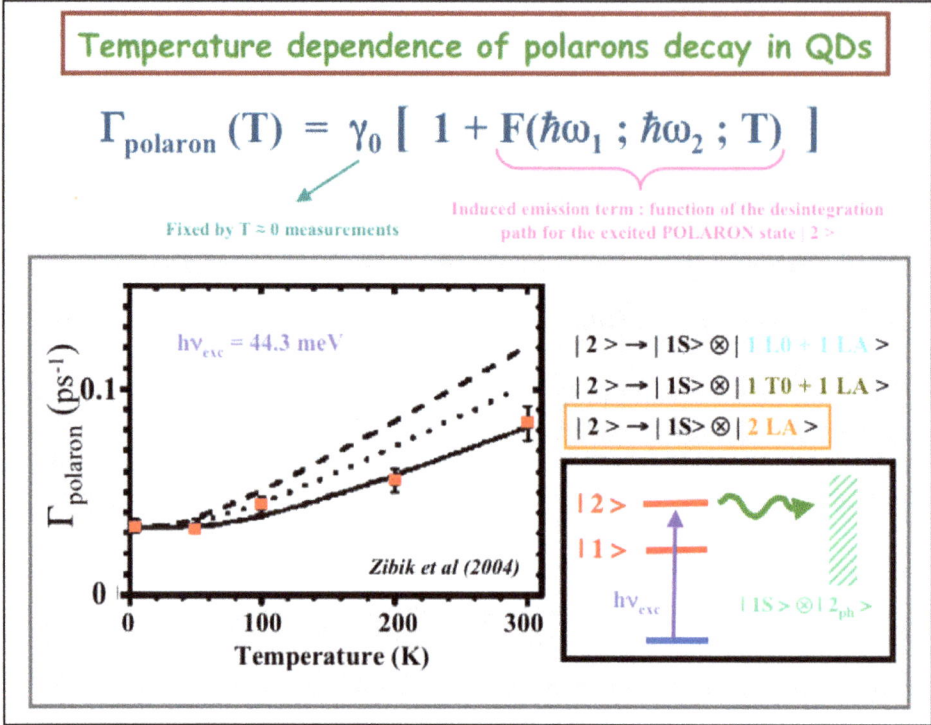

Figure 3.16. Left, symbols: variation with temperature of the measured relaxation rate for dots excited with energy $h\nu = 44.3$ meV. Left, curves: expected temperature variations for the three disintegration channels given on the right of the figure. Right: scheme of the experiment, where the upper polaron state $|2\rangle$ is excited. Adapted from Zibik *et al* (2004).

when the increasing polaron energy surpasses the upper limit of this path. This provides first evidence for the importance of this channel for dot polarons probed with an excitation energy $\hbar\omega_{LO} < h\nu = E_{|2\rangle} - E_{1S} < 53$ meV.

Further experimental evidence of the importance of the Klemens channel in this energy interval is presented in figure 3.16, which shows the temperature dependence of the decay time measured at constant excitation energy $h\nu = 44.3$ meV. The variation with temperature of the decay rate essentially reflects the population enhancement of the spontaneous two-phonon emission due to anharmonic coupling, which is expected to be different for the different channels. Indeed, the enhancement factor (see figure 3.16) is given by

$$1 + F(\hbar\omega_1; \hbar\omega_2; T) = \left[1 + n_B(\hbar\omega_1; T)\right]\left[1 + n_B(\hbar\omega_2; T)\right]$$
$$\approx 2 + n_B(\hbar\omega_1; T) + n_B(\hbar\omega_2; T), \tag{3.22}$$

where $n_B(\hbar\omega; T)$ is the Bose factor. As shown in figure 3.16, the observed variation at low temperature better follows that due to the 1LO $\rightarrow$ 2LA path: two emitted phonons of energy $h\nu/2$ and thus $F \approx 2\,n_B(h\nu/2; T)$. The more important variations of the paths involving one optical phonon (either LO or TO) come from the fact that

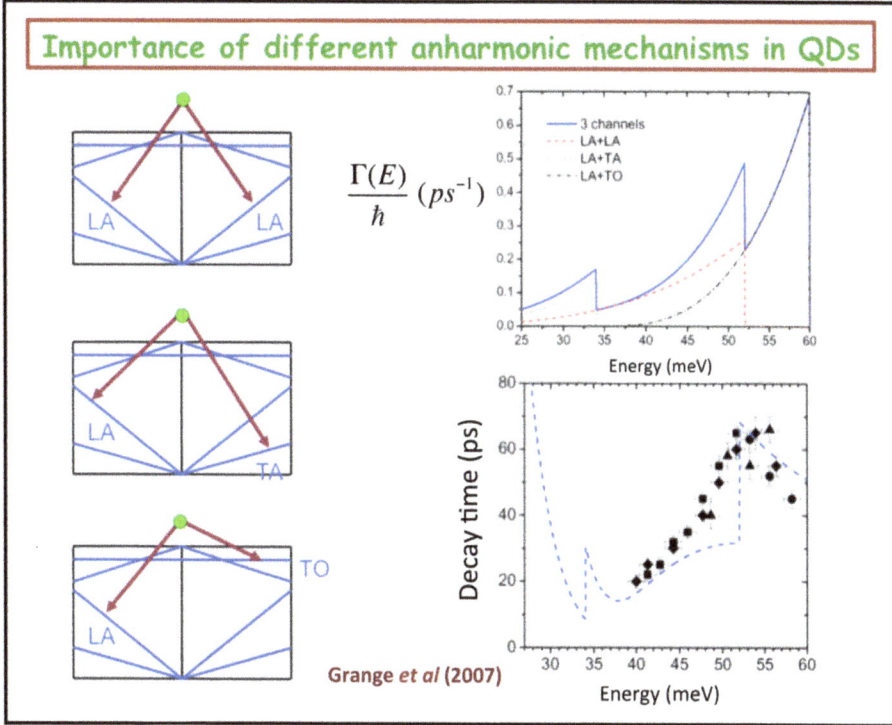

Figure 3.17. Left: sketches of the energy relaxation for the upper QD polaron with energy larger than E_{1S} + $\hbar\omega_{LO}$, for three channels: 2LA, 1LA + 1TA and 1LA + 1TO. Right: variations of the measured (symbols) and calculated (curves) mean relaxation times as functions of the excitation energy $h\nu$. Adapted from Zibik *et al* (2004) and Grange *et al* (2007).

the final LA phonons in these channels have considerably smaller energies than $h\nu/2$ (since the final optical phonon takes the largest part of the initial energy) and thus a thermal population larger than $2\, n_B(h\nu/2;\ T)$.

The last two experimental results highlight the importance of the Klemens channel for the relaxation of polarons with $\hbar\omega_{LO} < h\nu < 53$ meV. For higher energies, other paths should be considered, as discussed next.

The multi-channel problem was theoretically considered by Grange *et al* (2007), who used simple linear approximations for the phonons dispersions. Figure 3.17 presents their results for some important two-phonon channels: 1LA + 1LA (Klemens), 1LA + 1TA and 1LA + 1TO. The left panels are sketches of the corresponding relaxations, where the full circle points to the relaxed energy $E_{|2\rangle} - E_{1S} = \hbar\omega_1(\vec{Q}_1) + \hbar\omega_2(\vec{Q}_2)$ (see equation (3.20)), which is larger than $\hbar\omega_{LO}$ (or $h\nu > \hbar\omega_{LO}$). The upper-right panel shows the variations of $\Gamma_{\text{channel},1sp}(E = E_{Pol})$ with relaxed energy, for each channel, as well as the total value $\sum_{\text{channel}}\Gamma_{\text{channel},1sp}(E = E_{Pol})$. Finally, the lower-right-panel compares the calculated total mean time (inverse of the total rate obtained by multiplying the three-channel curve by the 1S weight $|\alpha(\Delta E_{pol})|^2$) with the low temperature data of Zibik *et al* (symbols). Note that the variation of the measured decay time around ≈ 53 meV is less abrupt than the calculated one. This is most probably due to the crude approximations for the phonon dispersions in the model. The results in figure 3.17 nonetheless clearly demonstrate that polaron instability is due to the coexistence of different relaxation channels, whose efficiencies significantly depend upon the energy of the QD transition. Indeed, for each channel one notes the presence of an edge discontinuity resulting from the finite width of two-phonon dispersions. Also, one can see the important energy-dependences, which trace back to different energy-dependent factors: the two-phonon joint dispersions inside the window; the strength of the anharmonic couplings'; and the weight of the one-phonon component of the polaron state. Only the last factor appears in the semiclassical description, leading to its inability to cope with the *decrease* of the decay time for increasing energy above ≈ 53 meV.

3.6.2 Relaxation of low-energy polarons

As shown in figure 3.17, Grange *et al* (2007) also considered the decay of low-energy polarons, i.e. of the state $|1\rangle$, which has energy below $E_{1S} + \hbar\omega_{LO}$ (or excitation energy below the optical phonon: $h\nu < \hbar\omega_{LO}$). The outcome of such calculations in a larger energy interval below the LO phonon is plotted as a full line in figure 3.18; note that the results are plotted in a log scale. Indeed, the calculations show a huge increase of the anharmonic decay time when the relaxing energy decreases far below $\hbar\omega_{LO}$. This increase is due to three factors that simultaneously decrease with increasing $|\Delta E|$: first, the one-phonon weight of the polaron state $|1\rangle$; second, the

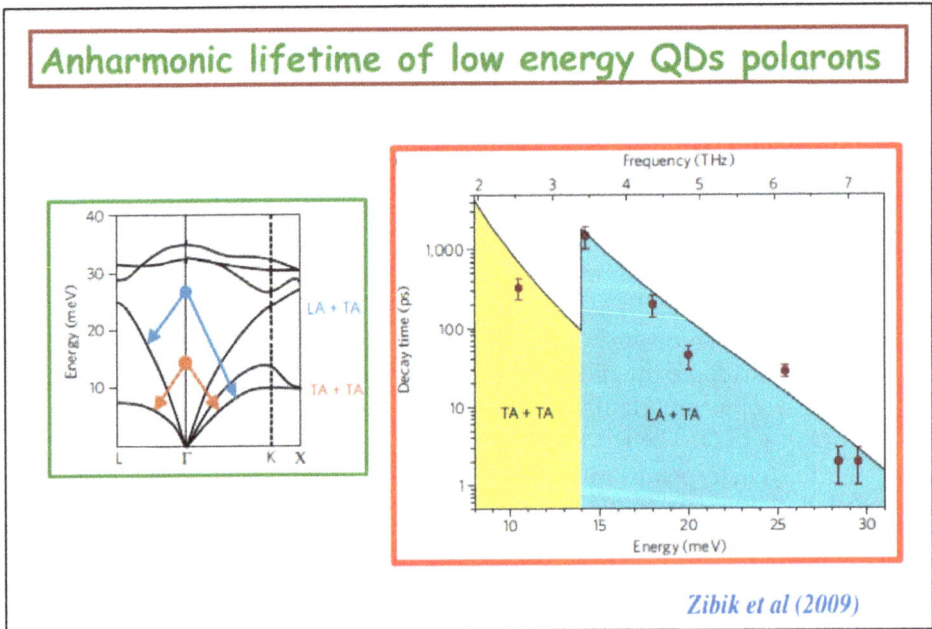

Figure 3.18. Variations of the measured (symbols) and calculated (curve) mean relaxation times as functions of the relaxation energy for low-energy polarons. From Zibik *et al* (2009).

shrinking of the final two-phonon density of states, since only low-energy acoustical phonons (LA and/or TA) can be emitted when the relaxing energy decreases (see the left panel in the figure); and finally, the strength of the anharmonic coupling itself (not discussed here) also decreases for the emission of two low-energy phonons.

It is worth pointing out that this huge variation, measured by Zibik *et al* (2009) and shown by the symbols in figure 3.18, corresponds to the observed mean decay times in dots of different S–P energetic separation. As indicated in this figure, the two principal disintegration channels are the 1LA + 1TA and the 1TA + 1TA ones, which are also sketched in the left panel of the figure for two different relaxing energies (full circles).

3.6.3 Conclusion

In conclusion, the existence on the one hand of a strong coupling between the confined electrons and the lattice LO vibrations, and on the other hand of the intrinsic anharmonic lifetime of such vibrations, provide an original theoretical framework to describe the population and energy relaxation in self-assembled QDs. The array of theoretical and experimental studies performed so far, for dot transitions both below and above the LO phonon energy, clearly corroborate this model. We finally stress the huge (by roughly three orders of magnitude) variation of the relaxation time in such dots (see figure 3.19). The dots with quasi-resonant

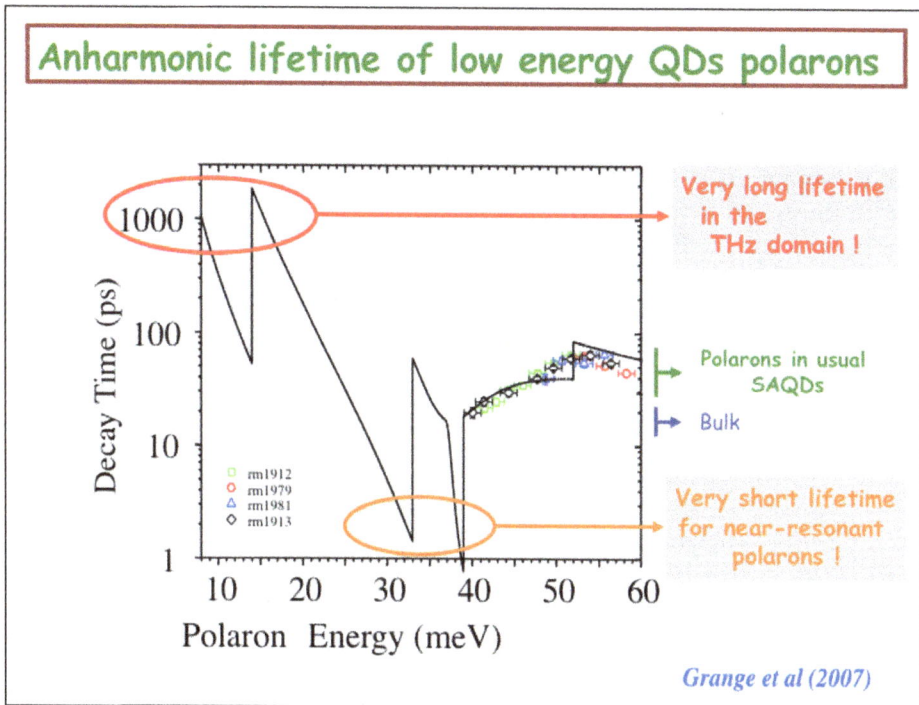

Figure 3.19. Variations of the measured (symbols) and calculated (curve) mean relaxation times as functions of the polaron relaxation energy. From Zibik *et al* (2004) and Grange *et al* (2007).

polaron states (very near $\hbar\omega_{LO}$) may have a relaxation time in the picosecond range, thus sizeably smaller than the one in bulk. Note that figure 3.19 also presents the contribution of the channel TA + LO, which introduces a very narrow additional relaxation window (not accessible in Zibik's experiments) right above the $\hbar\omega_{LO}$ excitation energy. Finally, 'THz' polarons have anharmonic decay in the nanosecond range, and are thus quasi-stable entities.

Both these results follow from the fact that the confinement in the dot leads to polaron states with energies that may greatly differ from the bulk value $\hbar\omega_{LO}$, the decay of which involve two-phonon channels of different energies and/or phonon components, which are definitely not accessible for bulk processes.

3.7 Intra-dot Auger relaxation

Let us consider now the relaxation processes triggered by the Coulomb coupling between two carriers bound in the dot: the intra-dot Auger relaxation. As discussed in chapter 1, excited states of a doubly occupied dot may be found to be energetically inside an ionization continuum. This is typically the case when the two occupied states are near the WL edge. In what follows, we shall not consider the modifications introduced by polaron couplings, and focus only on the role of the Coulomb interaction between confined carriers in the intra-dot Auger relaxation. We shall consider both electrons and one electron–hole pair.

For the case of electron–hole states, such a configuration was discussed in section 1.4.2. In practice, it may result from two consecutive captures by an empty dot after high-energy interband optical generation of free electron–hole pairs in the dot continuum, or by direct excitation of one electron–hole pair into excited conduction and valence band shells. These two situations are discussed later in the chapter.

Let us however start with the presentation of some general features of the Coulomb-triggered relaxation in the case of a dot hosting two electrons in its P shell, as illustrated in figure 3.20. The scheme in the right panel recalls some of the two-electron states (see figure 1.8 for a more complete scheme). In this figure, both the singlet and triplet states are resonances: they are inside the dissociation continuum where one electron is in the low-energy S level whereas the other is in the high-energy WL continuum. Thus, if we assume for instance that one singlet state is initially occupied (full circles in the right panel), then the system will irreversibly evolve towards the ensemble of states of the continuum. Let us note a few characteristics of this process.

(i) The irreversible evolution is triggered by the Coulomb potential, which couples the two two-electron configurations. This can be seen when we consider the two-electron eigenproblem $[H_1 + H_2 + V_C(1,2)] \Psi = E \Psi$, with $H_{1,2}$ the single-dot Hamiltonian for the two non-interacting electrons, and $V_C(1,2)$ their Coulomb coupling. If we search for solutions in the form (disregarding for a while spins and symmetrization; to be considered later)

$$\psi = \sum_{n,m} \alpha_{n,m} \psi_n(1) \psi_m(2), \tag{3.23}$$

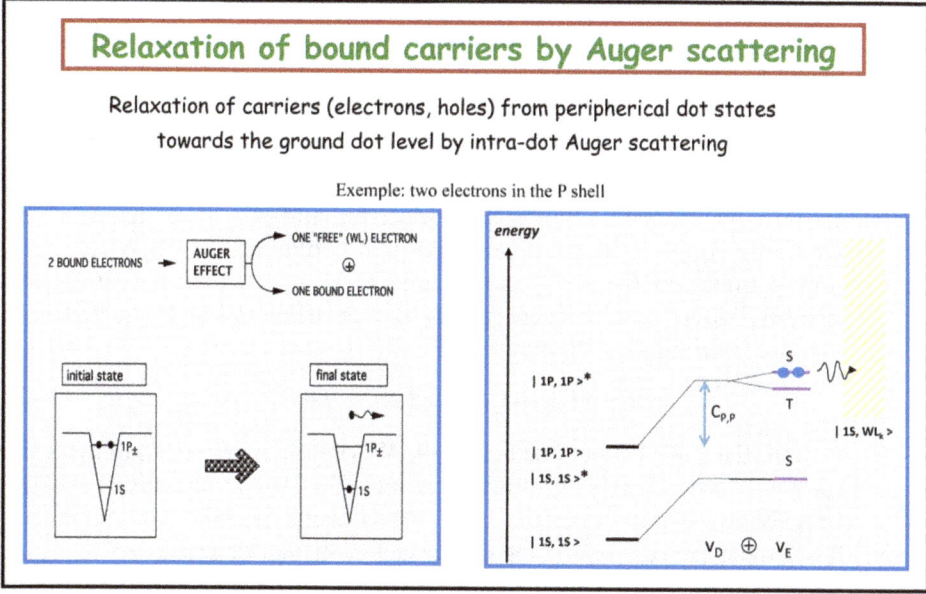

Figure 3.20. Scheme (left) and energy representation (right) of the intra-dot Auger mechanism allowing the relaxation of one electron from the P shell of a doubly occupied dot into the ground 1 s state, with ejection of the other electron into the continuum. Only some two-electron states are presented (see figure 1.8 for a more complete scheme).

where $H_1 \, \psi_n(1) = E_n \, \psi_n(1)$, we end up with the eigenmatrix problem

$$\left[E_n + E_m + V_{(n,m);(n,m)} - E \right] \alpha_{n,m} + \sum_{(N,M) \neq (n,m)} \alpha_{N,M} V_{(n,m);(N,M)} = 0, \quad (3.24a)$$

$$V_{(n,m);(N,M)} = \langle \psi_n(1)\psi_m(2) | \, V_C(1, 2) \, | \psi_N(1)\psi_M(2) \rangle. \quad (3.24b)$$

Noticing that ψ_n can be either a bound (discrete n-values) or a delocalized (continuum n-values) state results in three types of Coulomb-induced mixings: intra-discrete, intra-continuum and discrete-continuum. The intra-discrete diagonal (i.e. for $n = m = N = M$) couplings strongly shift the two-electron bound states, as described in chapter 1 (see figure 1.7); they are usually fully accounted for in the calculations. The non-diagonal terms (i.e. with one or more different states in the Coulombic matrix elements in equation (3.24b)) can in principle be diagonalized in the basis of the diagonal states (the so-called configuration interaction scheme), leading to inter-shell couplings that renormalize the energies of the stationary bound states. The intra-continuum couplings are often neglected, since they introduce smaller energy corrections and involve only high-energy states. The discrete-continuum coupling is of paramount importance in the relaxation issue, as discussed in the following.

(ii) The total energy is conserved, and the characteristic disintegration time can be evaluated with the aid of the Fermi golden rule. For the case represented in the figure 3.20, one has

$$E_{\text{fin}(2)} = E_{1S} + E_{WL} + E_k \geqslant E_{1S} + E_{WL}, \tag{3.25}$$

for the total energy of the final two-electron configuration, where $E_k \geqslant 0$ is the kinetic energy of the ejected electron and we have neglected any Coulomb coupling in the dissociated final configuration $|1S, WL_k\rangle$. If one approximates for the initial state energy $E_{\text{ini}(2)} \approx 2E_P + C_{P,P}$ with $C_{P,P}$, the repulsive direct Coulomb energy for the electrons in the P shell (denoted $v_D(e_{1P}, e_{1P})$ in figure 1.7), one gets

$$E_P - (E_{1S} - C_{P,P}) \geqslant E_{WL} - E_P, \tag{3.26}$$

i.e. that the P shell should be nearer the WL continuum edge than the S shell (diminished by $C_{P,P}$). This implies that intra-dot Auger relaxation is possible only for two-carrier states made of excited orbits near the continuum.

(iii) The Auger process conserves the singlet or triplet character of the initial state. This is clearly seem in equation (1.20b) for the case of different initial orbitals (say $n = P_+$ and $m = P_-$), since the Coulomb matrix element is proportional to $\delta_{\sigma,\Sigma}$. For identical initial orbitals ($n = m = P_+$ or P_-) only the $\sigma = +1$ combination is possible for the final state configuration, and we easily show that $J = K$ in this case.

(iv) The irreversible evolution produces the effective relaxation of one electron towards the ground state of the dot, with ejection of the other one into the WL or barrier continuum (left panel in figure 3.20). In the following, we focus on the characteristic time for such a relaxation process, evaluated with the Fermi golden rule for different initial states and assuming that the continuum states are initially empty and described by plane waves orthonormalized to the bound dot orbitals (see equation (1.17)).

Similar features hold for the case of one electron–hole pair, except (iii), which is replaced by the separate conservations of the electron and hole spins as well as of the total orbital angular momentum of the pair. Formally (and generally) speaking, one has in either case to deal with a Fano-like problem, whereby some discrete states are immersed in one (or various) dissociation continuum and coupled to it (to them) by $V_C(1,2)$. We shall focus initially on discrete–continuum coupling, which leads directly to an irreversible relaxation, but also discuss some additional aspects of the intra-discrete inter-level couplings later on in this chapter.

3.7.1 Two electrons

Figure 3.21 shows the variation of the Auger disintegration time for different initial configurations (singlet (S) or triplet (T), made of various orbitals ($P_{\pm 1}$, $D_{\pm 2}$)), as a function of the QD radius.

Note that each initial configuration contribûtes in a finite radius interval: the minimum radius is the one below which one or both of the initial orbitals are not

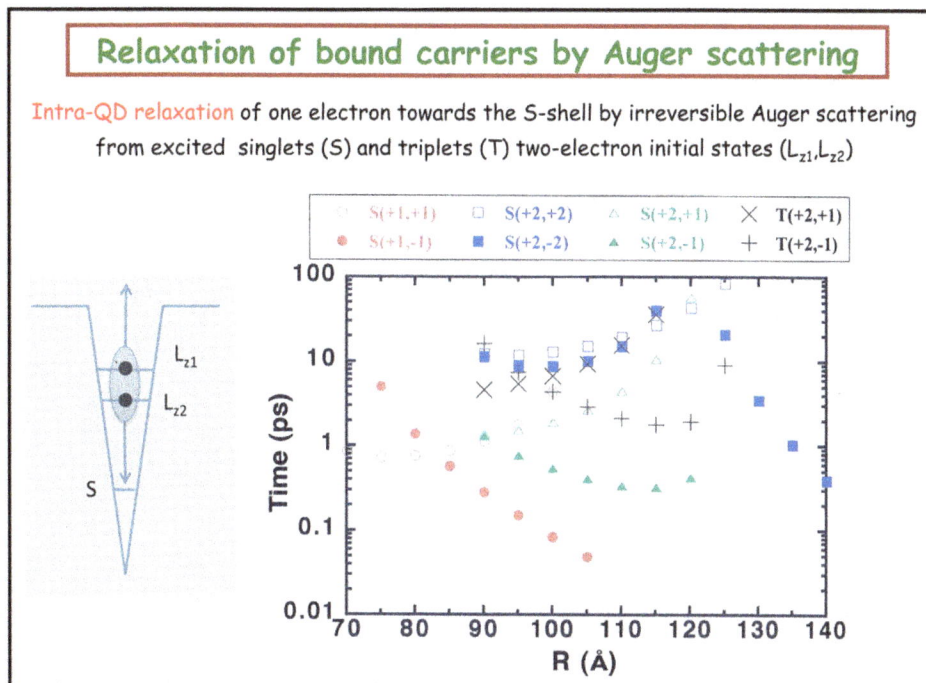

Figure 3.21. Scheme (left) and mean time (right) of the intra-dot Auger mechanism allowing the relaxation of one electron from one excited state of a doubly occupied dot into the ground 1s state, with ejection of the other electron into the continuum, for various initial singlet (S) or triplet (T) two-electron configurations made of states with different angular projections (L_{z1}, L_{z2}) along the growth axis. $L_z = +1$ for P_{+1},... Adapted from Ferreira and Bastard (1999).

bound, while for radii larger than the maximum one the initial state is too deep (it is no longer inside a dissociation continuum and therefore becomes stable).

The Auger scattering is in general an efficient relaxation mechanism: the relaxation time is typically below 10 ps. This efficiency reflects the strong Coulombic coupling between the carriers that are confined in the small dot region.

3.7.2 One electron–hole pair

The Auger mechanism is also efficient in the case of an electron–hole pair, as shown in figure 3.22 for the case where both carriers are initially in the P shell.

In the case of distinguishable particles, the exchange terms vanish: the Coulombic interaction conserves separately the total angular momentum and each of the two spins. Another difference with respect to the two-electron case is that one can point out the fate of each particle of the pair (relaxed or ejected). Note finally that this process is as efficient as the one for two electrons.

3.7.3 Other (temperature-dependent) intra-dot relaxation processes

In the case represented in figure 3.22, each carrier is initially in one excited state, such that the intrinsic irreversible Auger dissociation is well characterized. We shall

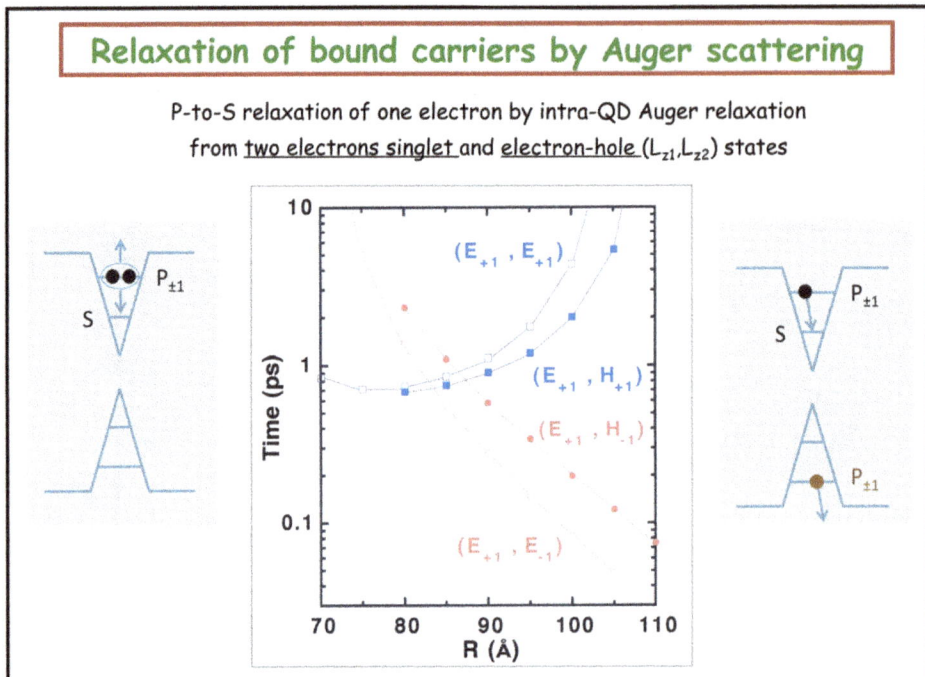

Figure 3.22. Schemes (left and right) and mean time (middle) of the intra-dot Auger mechanism allowing the relaxation of one electron from one excited P state of a doubly occupied dot into its ground 1S orbital, with ejection of either the other electron (left scheme) or one hole (right scheme) into the continuum, for different two-carrier configurations made of states in the P shells. The two-electron results correspond to the singlet ones in the previous figure. Adapted from Ferreira and Bastard (1999).

come back to this situation later in this paragraph. Let us for the moment consider the case where the pair energy is placed below any dissociation energy. In this case, it is worth considering in more detail the role of the intra-discrete Coulomb couplings mentioned in relation to equation (3.24). A full diagonalization of the terms $V_{(n,m);(N,M)}$ within the truncated sub-space span by the sole bound single-particle states leads to mixed stationary solutions, and corresponds to the so-called configuration inter-action scheme to describe the two-particle states of the dot. As a consequence, such intra-discrete mixings cannot *a priori* produce an irreversible evolution of one of the carriers towards its ground state. Such a relaxation can nevertheless occur if the bound one-carrier states are actually unstable. This is the case in the presence of polaron couplings and anharmonic effects, as previously discussed in this chapter. This is also the case if the dot possesses many bound states, energetically near enough so as to be efficiently coupled by acoustic phonon emissions and absorp-tions. This case has been initially discussed by Efros *et al* (1995) and applied to the energy relaxation in nanometric-sized QDs (or nanocrystals; not discussed in this book). Figure 3.23 shows the results of calculations of Narvaez *et al* (2006) of the mean time in self-assembled dots for the case where one electron initially in the P shell relaxes down to the S level because of its Coulomb interaction with one hole

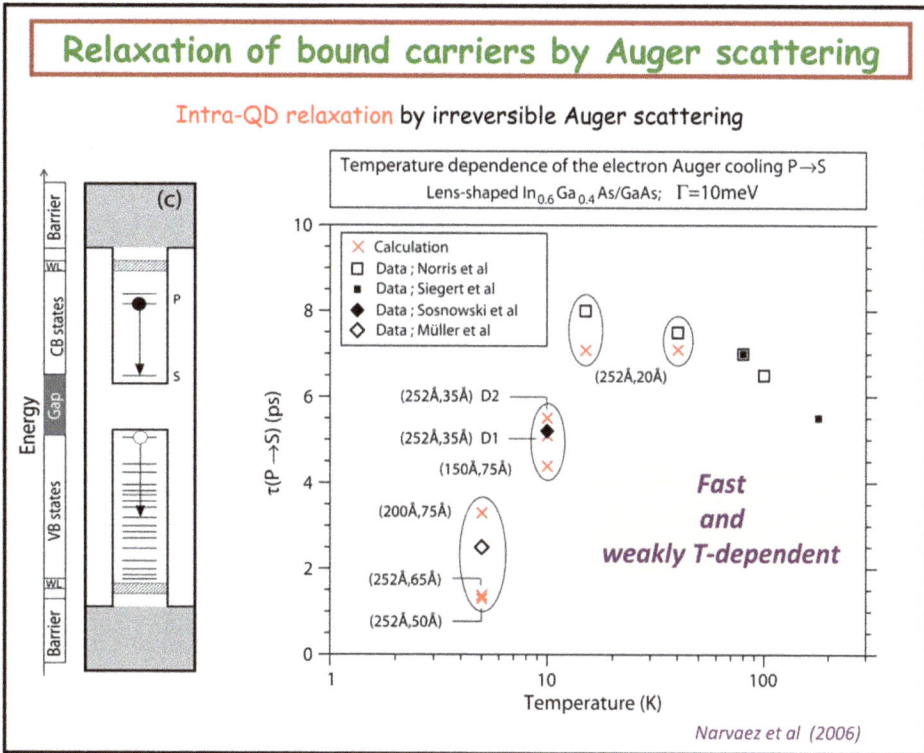

Figure 3.23. Scheme (left) and mean time (right) of the intra-dot Auger mechanism allowing the relaxation of one electron from one excited P state of a doubly occupied dot into its ground orbital, with excitation of one hole from its ground state towards an excited level. Hole level broadening $\Gamma = 10$ meV. Adapted from Narvaez *et al* (2006).

whose states are broadened. In their calculations, an arbitrary phenomenological broadening ($\Gamma = 10$ meV) introduces relaxation and thermally activated excitation channels for the holes. It means that the electron can irreversibly transfer its initial energy to the hole (initially in its ground state in the calculations, as depicted in the left panel of figure 3.23. The relaxation mean time becomes temperature-dependent, as shown in the figure. Like in the previous figures, the calculated electron relaxation times are short (in the range of a few picoseconds). The symbols in the figure represent the outcome of various experiments (for details, see references quoted in Narvaez *et al* (2006)).

Let us finally come back to the case of one electron–hole pair with both carriers in excited states (as shown in figure 3.22). As discussed in section 1.4.2, the pair spectrum presents resonances (see figure 1.10). The latter have been studied experimentally and their spectral width measured as a function of temperature. The principle of the experiments on a single QD is shown in figure 3.24.

In a first step, one measures the photoluminescence (PL) spectrum of the isolated dot under excitation at high energy (i.e. in the WL of the dot; lower-left panel in figure 3.24). In this experiment, free electron–hole pairs photoexcited in the WL continuum are subsequently captured by the dot and relax down to its

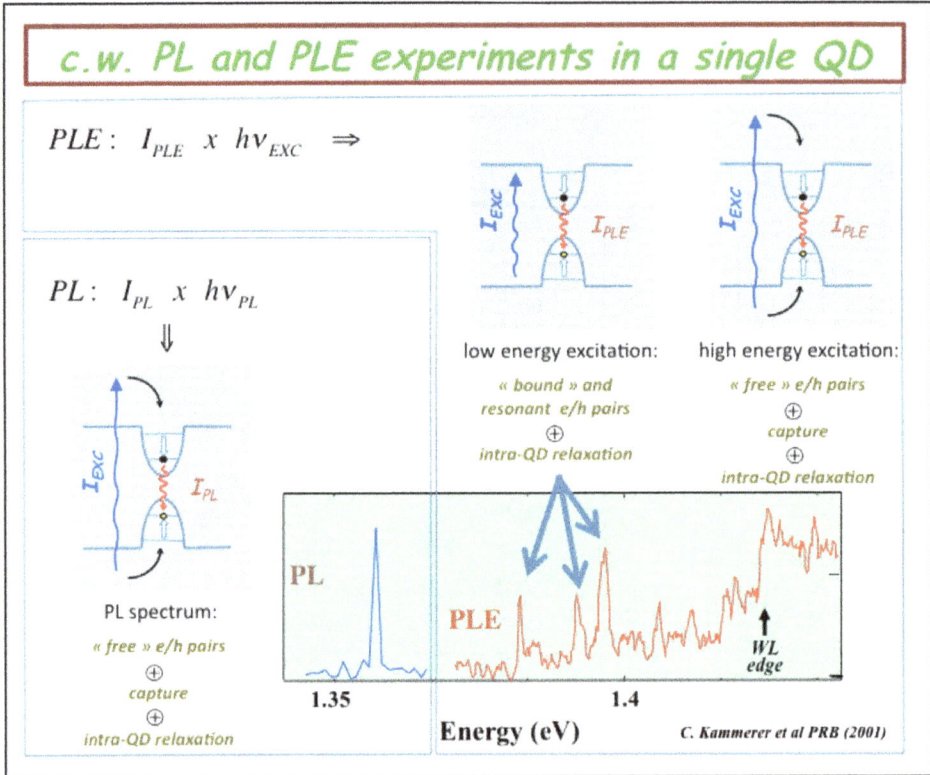

Figure 3.24. Schematic representations and measured intensities of the photoluminescence (PL; lower-left) and excitation of the photoluminescence (PLE; right) spectra of an isolated QD. Adapted from Kammerer *et al* (2001).

ground interband state, from where a radiative recombination occurs. The photon emission spectrum allows the energy E_F of the low interband transition to be determined (≈ 1.36 eV in the figure). In a second step, the detection energy is fixed at E_F and one collects the photoluminescence excitation (PLE) spectrum; namely, the variation of the intensity of the photon emission of the dot in its ground interband state as a function of the exciting photon energy $h\nu_{EXC}$ (with $h\nu_{EXC} > E_F$; right panel in figure 3.24).

The PLE spectrum displays a series of peaks (some of them pointed by the arrows), which are spectrally localized in between the ground dot level and the edge of the WL continuum: $E_F < E < E_{WL}$; these transitions are due to the optical excitation of either the excited bound states or the resonances of the electron–hole pairs.

The lower-left panel in figure 3.25 shows that the measured width $\Gamma(T)$ of one such resonance peak has a finite value at very low temperatures ($\Gamma(0) \approx 0.3$ meV in the present case) and then increases roughly linearly with increasing temperature up to about 50 K.

The value of $\Gamma(0)$ and the linear increase of $\Gamma(T)$ are to be associated with various mechanisms affecting the resonant pair states. To understand this point, the right

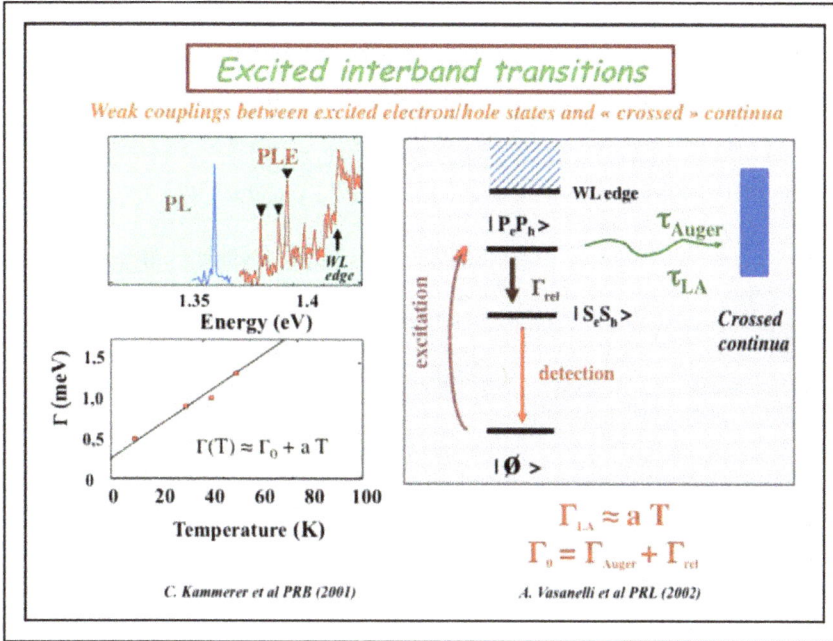

Figure 3.25. Upper left: PL and PLE spectra (same as in figure 3.24). Lower left: measured width of one resonance peak (marked by arrows in the upper-left panel). Right: scheme of the photoexcitation and energy relaxation paths before detection at the energy of the ground dot level. Adapted from Kammerer *et al* (2001) and Vasanelli *et al* (2002).

panel of the figure shows the various states involved in the experiments: from the ground $|\varnothing\rangle$ state (representing the initially empty dot), one electron–hole pair is photoinjected in a resonant level ($|P_e, P_h\rangle$ in the scheme), from which the pair either relaxes down to the fundamental $|S_e, S_h\rangle$ state, with a rate Γ_{rel} (due, for instance, to the polaron couplings acting on the electron–hole pair), or makes an irreversible transition towards the crossed continuum. Two couplings can trigger the latter: the Coulombic one and the interaction with low-energy acoustical phonons. The first corresponds to the intra-dot Auger processes, which is at lowest order temperature-independent. The second, in contrast, increases rapidly with temperature and, above $T = 10$ K or so, presents a linear variation with temperature (which simply reflects the increasing population of low-energy phonons involved in the scattering process). The calculation (not detailed in this work) of the slope 'a' of the linear variation $\Gamma_{LA} = aT$ is in good agreement with measurements, thereby corroborating the interpretation of the temperature enhancement of the peak broadening in terms of a Fano-like dissociation assisted by the acoustical phonon couplings. The measured low-temperature broadening is associated with the polaron and Auger contributions: $\Gamma_0 = \Gamma_{rel} + \Gamma_{Auger}$; the weight of each contribution cannot however be experimentally separated. Its value nonetheless indicates an efficient intra-dot relaxation even at low temperatures.

References

Benisty H, Sottomayor-Torres C M and Weisbuch C 1991 Intrinsic mechanism for the poor luminescence properties of quantum-box systems *Phys. Rev. B* **44** R10945

Bockelmann U and Bastard G 1990 Phonon scattering and energy relaxation in two-, one-, and zero-dimensional electron gases *Phys. Rev. B* **42** 8947

Bockelmann U and Egeler T 1992 Energy relaxation in quantum dots by means of Auger processes *Phys. Rev. B* **46** R15574

Efros Al L, Kharchenko V A and Rosen M 1995 Breaking the phonon bottleneck in nanometer quantum dots: Role of Auger-like Processes *Solid State Commun.* **93** 281

Ferreira R and Bastard G 1999 Phonon-assisted capture and intra-dot Auger relaxation in quantum dots *Appl. Phys. Lett.* **74** 2818

Grange T, Ferreira R and Bastard G 2007 Polaron relaxation in self-assembled quantum dots: Breakdown of the semiclassical model *Phys. Rev. B* **76** 241304(R)

Kammerer C, Cassabois G, Voisin C, Delalande C, Roussignol Ph, Lemaître A and Gérard J M 2001 Efficient acoustical phonon broadening in single self-assembled InAs/GaAs quantum dots *Phys. Rev. B* **65** 033313

Inoshita T and Sakaki H 1992 Electron relaxation in a quantum dot : significance of multiphonon processes *Phys. Rev. B* **46** 7260

Li X-Q and Arakawa Y 1998 Anharmonic decay of confined optical phonons in quantum dots *Phys. Rev. B* **57** 12285

Li X-Q and Arakawa Y 1997 Ultrafast energy relaxation in quantum dots through defect states: A lattice-relaxation approach *Phys. Rev. B* **56** 10423

Li X-Q, Nakayama H and Arakawa Y 1999 Phonon bottleneck in quantum dots: Role of lifetime of the confined phonons *Phys. Rev. B* **59** 5069

Narvaez G A, Bester G and Zunger A 2006 Carrier relaxation mechanisms in self-assembled (In, Ga)As/GaAs quantum dots: Efficient P->S Auger relaxation of electrons *Phys. Rev. B* **74** 075403

Rossi F and Zanardi P (ed) 1995 *Semiconductor Macroatoms* (London: Imperial College Press)

Sauvage S, Boucaud P, Lobo R P S M, Bras F, Fishman G, Prazeres R, Glotin F, Ortega J M and Gérard J-M 2002 Long Polaron Lifetime in InAs/GaAs Self-Assembled Quantum Dots *Phys. Rev. Lett.* **88** 177402

Vasanelli A, Ferreira R and Bastard G 2002 Continuous absorption background and decoherence in semiconductor quantum dots *Phys. Rev. Lett.* **89** 216804

Verzelen O, Ferreira R and Bastard G 2000 Polaron lifetime and energy relaxation in semiconductor quantum dots *Phys. Rev. B* **62** R4809

Zibik E A *et al* 2009 Long lifetimes of quantum-dot intersublevel transitions in the terahertz range *Nature Mater.* **8** 803

Zibik E A *et al* 2004 Intraband relaxation via polaron decay in InAs self-assembled quantum dots *Phys. Rev. B* **70** 161305(R)

www.ingramcontent.com/pod-product-compliance
Lightning Source LLC
Chambersburg PA
CBHW081549220326

41598CB00036B/6613